AF333200

CALCIUM CHANNELS

PROPERTIES, FUNCTIONS AND REGULATION

CEREBROVASCULAR RESEARCH AND DISORDERS

Additional books in this series can be found on Nova's website under the Series tab.

Additional E-books in this series can be found on Nova's website under the E-book tab.

PHYSIOLOGY - LABORATORY AND CLINICAL RESEARCH

Additional books in this series can be found on Nova's website under the Series tab.

Additional E-books in this series can be found on Nova's website under the E-book tab.

CALCIUM CHANNELS

PROPERTIES, FUNCTIONS AND REGULATION

MARK R. FIGGINS

EDITOR

Nova Science Publishers, Inc.
New York

Copyright © 2012 by Nova Science Publishers, Inc.

For permission to use material from this book please contact us:
Telephone 631-231-7269; Fax 631-231-8175
Web Site: http://www.novapublishers.com

NOTICE TO THE READER

The Publisher has taken reasonable care in the preparation of this book, but makes no expressed or implied warranty of any kind and assumes no responsibility for any errors or omissions. No liability is assumed for incidental or consequential damages in connection with or arising out of information contained in this book. The Publisher shall not be liable for any special, consequential, or exemplary damages resulting, in whole or in part, from the readers' use of, or reliance upon, this material. Any parts of this book based on government reports are so indicated and copyright is claimed for those parts to the extent applicable to compilations of such works.

Independent verification should be sought for any data, advice or recommendations contained in this book. In addition, no responsibility is assumed by the publisher for any injury and/or damage to persons or property arising from any methods, products, instructions, ideas or otherwise contained in this publication.

This publication is designed to provide accurate and authoritative information with regard to the subject matter covered herein. It is sold with the clear understanding that the Publisher is not engaged in rendering legal or any other professional services. If legal or any other expert assistance is required, the services of a competent person should be sought. FROM A DECLARATION OF PARTICIPANTS JOINTLY ADOPTED BY A COMMITTEE OF THE AMERICAN BAR ASSOCIATION AND A COMMITTEE OF PUBLISHERS.

Additional color graphics may be available in the e-book version of this book.

LIBRARY OF CONGRESS CATALOGING-IN-PUBLICATION DATA

Calcium channels : properties, functions, and regulation / [edited by] Mark R. Figgins.
    p. ; cm.
Includes bibliographical references and index.
ISBN 978-1-61470-232-0 (hardcover : alk. paper) 1. Calcium channels. I. Figgins, Mark R.
[DNLM: 1. Calcium Channels. QU 55.7]
QP535.C2C2626 2011
572.516--dc23
           2011020242

Published by Nova Science Publishers, Inc. † New York

Contents

Preface   vii

Chapter 1 The Distribution of Calcium Channel Cav1.3 in the Central Nervous System and Its Functions in Relation to Motor Control   1
Mengliang Zhang, Hans Hultborn and Natalya Sukiasyan

Chapter 2 The Distinct Signaling Mechanisms of N-Type and L - Type Calcium Channels in Cardiac Hypertrophy and Failure   49
N. Petrashevskaya, E. Kobrinsky and A. Schwartz

Chapter 3 The Role of Voltage Gated Calcium Channels in the Developing Vertebrate Retina   87
Elizabeth A. MacMurray and Margaret S. Saha

Chapter 4 Unraveling The Multiple Functions of the $\alpha2/\delta$ Subunits of Calcium Channels   109
Jesús García

Chapter 5 Mesoscopic Simulation of Subcellular Calcium Microdomains and Calcium-Regulated Calcium Channels   121
Nicolas Wieder, Rainer H. A. Fink and Frederic von Wegner

Chapter 6 Lethargic Mice: An Interesting Animal Model to Investigate Voltage-Gated Calcium Channel ß_4 Subunit Function   137
Katell Fablet and Michel De Waard

Chapter 7 Activation Mechanisms of Store-Operated Calcium Channels in Human Neutrophils   151
Costantino Salerno and Carlo Crifò

Chapter 8 Membrane Transport of Toxic Metals by Ionic Mimicry in Mammalian Cells: Where Do Calcium Channels Fit In?   163
Carla Marchetti

Index   185

PREFACE

Calcium channels are ion channels which display selective permeability to calcium ions and are sometimes synonymous as voltage-dependent calcium channels. In this book, the authors present research from across the globe in the study of the properties, functions and regulation of calcium channels. Topics discussed include the distribution of calcium channel Cav1.3 in the central nervous system; the signaling mechanisms of N-type and L-type calcium channels in cardiac hypertrophy and failure; voltage gated calcium channels in the developing vertebrate retina and subcellular calcium dynamics in femtoliter volumes based on the Gillepsie algorithm.

Chapter 1 - Cav1.3 is one of the voltage-gated L-type calcium channels, which plays an important role in controlling the activity in many types of excitable tissues. It controls, for example, the pacemaker function of the sinoatrial node, the auditory signal transduction at the level of the hair cells as well as the gain of the input-output function of motoneurons. This chapter presents the recent advances in the study of the distribution of Cav1.3 channels across the central nervous system, with a special focus on the spinal cord across several animal species including the cat, rat, mouse and turtle. The functions of the channel are discussed in relation to motor control, not least to the motoneurons themselves.

In the spinal cord Cav1.3 has been found to be expressed in all of the above mentioned animal species; however its detailed distribution patterns have been shown to vary across species. In cats, rats and mice Cav1.3 was expressed in all parts of the spinal gray matter although in cats the expression in the dorsal horn was significantly less than in the ventral horn in comparison with the other two species. Furthermore, Cav1.3 immunoreactivity was significantly higher in ventral horn motoneurons than in neurons located in dorsal horn and intermediate zone. The spatial distribution of the channel immunoreactivity along the soma-dendritic axes seems also different in the different species. In cats and rats the neuronal somata were densely immunolabeled, although a clear immunoreactivity could also be seen in their primary and secondary dendritic branches. In mice Cav1.3 immunoreactivity was revealed to be mainly distributed in the dentrites, especially the distal dendrites. In turtles Cav1.3 immunoreactivity was found to be predominantly located on the motoneuronal dendrites, including their proximal and the distal parts, although it could also be seen on the cell somata. However, using antibodies from different sources different immunolabeling patterns have been produced, which raises the possibility that different antibodies may be specific for different splice variants that may have a specific soma-dendritic distribution. The distribution of Cav1.3 channels in the brain was mainly examined in rats and it was shown that the channel was expressed in the olfactory bulb, cerebral cortex, hippocampus, basal ganglia, diencephalon, cerebellum and brain stem. In the brain stem Cav1.3 was expressed extensively in different regions including both motor and sensory nuclei with the neurons in the former group being more densely immunolabeled than the later group. Some nuclei, which contain monoaminergic neurons, for example the locus coeruleus in rats, were also densely immunolabeled.

Due to its low activation threshold, it is strongly believed that Cav1.3 is one of the main channel types that mediate the persistent inward currents (PICs) in spinal motoneurons. Although it is still under debate as to where the PICs originate along the soma-dendritic membrane, physiological evidence suggests that these currents mainly originate from dendrites, although it can also originate from neuronal somata. In motoneurons the PICs could evoke all-or-none plateau potentials following injection of brief current pulses. With normal synaptic excitation there is evidence for a gradual recruitment of the PICs. Since the Cav1.3 channels in motoneurons are facilitated by monoaminergic innervation it has been suggested that the PICs mediated by Cav1.3 may serve as a controlled amplifier of synaptic excitation to motoneurons.

Chapter 2 - Pathologic cardiac hypertrophy is an independent risk factor for myocardial infarction, arrhythmias, and subsequent heart failure. Hypertrophic remodeling and associated hypertrophic growth, fibrosis, cavity dilatation and electrophysiological remodeling occur in response to hemodynamic stresses such as hypertension, myocardial infarction, and valvular insufficiencies. The prevalence of hypertrophy and heart failure is increasing rapidly worldwide, and effective treatments remain elusive. Cardiac hypertrophy and failure is a complex syndrome, which involves neuro-endocrine stimulation and activation of Ca^{2+} signaling pathways predominantly within cellular microdomains to induce pathologic pro-growth signaling cascades. Developing new therapies to target pathological remodeling of the hearth should synergistically target excessive sympathetic activation and molecular mechanisms of ventricular remodeling

New complex heart failure therapies are based on N-type and L- type calcium channel blockers and Ca-dependent signaling pathways modulators that target sympathetic hyper-activation and Ca^{2+}dependent proximal signaling with intent to prevent, halt, or reverse the progression of disease.

Chapter 3 - While the role of voltage-gated calcium channels (VGCCs) in the adult retina is well established, their role in development and in the embryonic retina has received less attention. However, in the developing retina, the diversity of calcium channels, particularly the L-type and T-type, contributes to cell proliferation, differentiation, and synaptogenesis. Additionally, VGCCs may contribute to multiple forms of intracellular calcium activity which regulates developmental processes. L-type channels are functional prior to synapse formation and may regulate optic cup formation. Both T-type and L-type channels are involved in retinal cell proliferation and are possibly involved in the proliferative mechanism known as interkinetic nuclear migration. T-type channels may contribute to retinal progenitor cell differentiation and are typically downregulated during late development. Transient high densities of the T-type channels are associated with undifferentiated retinal progenitors and undifferentiated retinoblastoma cell lines. L-type channels are necessary for retinal synapse formation and establishing retinogeniculate connections via spontaneous retinal waves. The necessity of L-type channels in synaptogenesis is exhibited by mutations in the Cav1.4 subunit, an essential component for photoreceptor ribbon synapses. Further research on calcium channels may elucidate their role in important aspects of retinal development as well as their potential as therapeutic targets for cell repair.

Chapter 4 - The $\alpha2/\delta$ subunits form part of voltage-dependent calcium channels and are considered accessory proteins. Initial studies of $\alpha2/\delta$ subunits revealed that they modify the voltage dependence, amplitude, and kinetics of calcium currents. $\alpha2/\delta$ subunits have also been identified as the receptor for gabapentin and pregabalin, drugs clinically used in the treatment of several neuropathic disorders. In addition to mediating therapeutical mechanisms, the involvement of $\alpha2/\delta$ subunits in diseases has been demonstrated in mouse models with mutations of the protein or alterations in expression levels. The structure of $\alpha2/\delta$ subunits and their pattern of expression suggested that these proteins may be participating in cellular functions separate from the action of calcium channels. Recently, it has been shown that $\alpha2/\delta$ subunits are more than just accessory subunits of calcium channels and that they are important for the trafficking of channels, attachment and migration of myoblasts, synaptogenesis, and other cellular processes. It is clear that $\alpha2/\delta$ subunits play many and important roles in the normal functioning of cells. This chapter reviews the influence of $\alpha2/\delta$ subunit in setting of the biophysical properties of calcium channels, the connection of $\alpha2/\delta$ with diseases, and the functions of $\alpha2/\delta$ that are now being unraveled.

Chapter 5 - In this chapter the authors present a stochastically exact, full Markovian model of localized, subcellular calcium dynamics in femtoliter volumes based on the Gillespie algorithm. They show that a stochastic approach is necessary to accurately describe the events that constitute the relevant steps of subcellular signal processing in many cell types, and especially in excitable tissues. They also discuss how different buffer systems shape the stochastic properties of the local calcium signal on several time scales. Using the chemical Langevin equation (CLE) as an approximation to the solution computed with Gillespie's algorithm, the local calcium signal can be treated as a colored noise input to downstream calcium-dependent signalling cascades. Importantly, the colored noise approximation achieved by the CLE approach allows for accelerated simulations. To exemplify the procedure the authors will include the calcium regulated IP$_3$R calcium channel in the simulations to produce a local positive feedback loop. This system represents the basic mechanism responsible for elementary calcium release events (blips, puffs, sparks) as well as for global calcium waves and oscillations. The inherent stochasticity of localized calcium dependent signalling is illustrated by including calmodulin and calmodulin dependent protein kinase IIa (CaMKIIa) in the model and it is shown that individual time courses of activated CaMKIIa display a large variability that has to be taken into account when investigating downstream signalling events such as CaMKIIa dependent protein phosphorylation-dephosphorylation switches.

Chapter 6 - *Lethargic* (*lh*) mice are spontaneous mutant mice presenting a loss of β_4 subunit of voltage-gated calcium channels (VGCC) due to a mutation in the gene encoding for β_4, Cacnb4. These *lh* mice are characterized by a complex phenotype with notably severe neurobehavioral defects. They comprise gait ataxia, paroxysmal dyskinesia, hypokinetic behavior and absence epilepsy seizures. The epileptopathogenesis has been extensively investigated in this mutant strain. Disturbances of the GABA$_B$ receptor system with the involvement of T-type VGCC in the thalamus may explain the appearance of absence epilepsy. Considering the role of β subunits in VGCC assembly, targeting and modulation of voltage-sensitive parameters, loss of β_4 has severe consequences for VGCC functions. Data from *lh* mice gave insight into an interesting phenomenom called "subunit reshuffling". Indeed, a compensation mechanism by β subunits has been observed in certain brain regions of *lh* mice. Far from being complete, "subunit reshuffling" is evidence of neuronal plasticity that may attenuate the severe defects caused by the absence of β_4 in *lh* mice. Where compensation is not possible, the absence of β_4 highlights additional roles of β_4 possibly in gene regulation.

Chapter 7 - Calcium mobilization plays an important role in the regulation of superoxide anion secretion by neutrophils. Intracellular calcium increase is predominantly a result of calcium influx from extracellular media through the opening of calcium-permeable channels, which is subsequent to the emptying of intracellular stores. This capacitative mechanism, referred as store-operated calcium entry (SOCE), implies that depletion of agonist-sensitive intracellular calcium stores generates a secondary signal that promotes plasma membrane calcium influx. Accumulating evidence indicates that three protein families (TRPC, STIM, and Orai) play obligatory roles in the activation of this pathway by forming a ternary heterologous complex on the plasma membrane. This complex might mediate communication between the endoplasmic reticulum and the plasma membrane, perhaps by facilitating coupling between TRPC and inositol 3,4,5-trisposphate receptors. STIM1 behaves as a sensor of Ca^{2+} level in the endoplasmic reticulum, while Orai 1, by interacting with TRPC and STIM1, might act as a regulatory subunit that operates the transduction of the signals to the calcium-permeable channels on the plasma membrane. The role of the microtubule network in the regulation of SOCE is still obscure. Experiments performed with microtubule inhibitors gave rise to contradictory results, since these compounds induced partial depletion of Ca^{2+} stores and influx of bivalent cations from extracellular medium in the cytosol, but at the same time they inhibited SOCE triggered by agonists that are known to deplete Ca^{2+} from the endoplasmic reticulum.

Chapter 8 - Cellular membranes are basically impermeable to ions and have developed specific pathways (transporters, channels or pumps) to facilitate metal translocation. These physiological carriers are not ideally selective and their specificity spectrum may include xenobiotic species, such as toxic metals whose availability in the environment has increased enormously with the onset of the industrial era. Competition between divalent endogenous and noxious metal ions at specific sites on membrane transport proteins and enzymes is referred to as "ionic mimicry".

In this chapter the author will present some studies on the permeation mechanisms through mammalian cell plasma membranes of lead (Pb) and cadmium (Cd), two metals whose toxicity has been linked to their putative ability to mimic calcium (Ca) and zinc (Zn) at specific binding sites. Both metals can permeate through mammalian cell membranes taking advantage of different Ca channels, but, while Cd appears to take advantage mainly of the same pathways as Ca, Pb is also rapidly taken up in different cell types by passive transport systems that are distinct from Ca channels and independent of specific stimuli.

To further elucidate the role of voltage-dependent Ca channels (VDCC) in Cd uptake, they compared the effect of this metal in two Chinese hamster ovary (CHO) cell lines, a wild type and modified cell line, which was permanently transfected with an L-type VDCC. Both cultures were subjected to brief (30-60 min) exposure to 50-100 µM Cd in serum-free culture medium. Cell death was evident after 18-24 h with comparable features in both cell lines. Although VDCC represent a pathway of Cd entry and participate in Cd-induced toxicity, as demonstrated by the effect of DHP modifiers, expression of L-type Ca channels is not sufficient to modify Cd accumulation and sensitivity to a toxicologically significant extent. This study confirmed that both Cd and Pb can take advantage of VDCC to permeate the membrane, but these transport proteins are not the only, and frequently not the most important, pathways of permeation.

Chapter 1

THE DISTRIBUTION OF CALCIUM CHANNEL CAV1.3 IN THE CENTRAL NERVOUS SYSTEM AND ITS FUNCTIONS IN RELATION TO MOTOR CONTROL

Mengliang Zhang[1,2], Hans Hultborn[1] and Natalya Sukiasyan[1]*
[1] Department of Neuroscience and Pharmacology the Panum Institute,
University of Copenhagen, Copenhagen, Denmark
[2] Department of Exercise and Sport Sciences, University of Copenhagen,
Copenhagen, Denmark

Abstract

Cav1.3 is one of the voltage-gated L-type calcium channels, which plays an important role in controlling the activity in many types of excitable tissues. It controls, for example, the pacemaker function of the sinoatrial node, the auditory signal transduction at the level of the hair cells as well as the gain of the input-output function of motoneurons. This chapter presents the recent advances in the study of the distribution of Cav1.3 channels across the central nervous system, with a special focus on the spinal cord across several animal species including the cat, rat, mouse and turtle. The functions of the channel are discussed in relation to motor control, not least to the motoneurons themselves.

In the spinal cord Cav1.3 has been found to be expressed in all of the above mentioned animal species; however its detailed distribution patterns have been shown to vary across species. In cats, rats and mice Cav1.3 was expressed in all parts of the spinal gray matter although in cats the expression in the dorsal horn was significantly less than in the ventral horn in comparison with the other two species. Furthermore, Cav1.3 immunoreactivity was significantly higher in ventral horn motoneurons than in neurons located in dorsal horn and intermediate zone. The spatial distribution of the channel immunoreactivity along the soma-dendritic axes seems also different in the different species. In cats and rats the neuronal somata were densely immunolabeled, although a clear immunoreactivity could also be seen in their primary and secondary dendritic

---

* E-mail address: mzhang@sund.ku.dk

branches. In mice Cav1.3 immunoreactivity was revealed to be mainly distributed in the dentrites, especially the distal dendrites. In turtles Cav1.3 immunoreactivity was found to be predominantly located on the motoneuronal dendrites, including their proximal and the distal parts, although it could also be seen on the cell somata. However, using antibodies from different sources different immunolabeling patterns have been produced, which raises the possibility that different antibodies may be specific for different splice variants that may have a specific soma-dendritic distribution. The distribution of Cav1.3 channels in the brain was mainly examined in rats and it was shown that the channel was expressed in the olfactory bulb, cerebral cortex, hippocampus, basal ganglia, diencephalon, cerebellum and brain stem. In the brain stem Cav1.3 was expressed extensively in different regions including both motor and sensory nuclei with the neurons in the former group being more densely immunolabeled than the later group. Some nuclei, which contain monoaminergic neurons, for example the locus coeruleus in rats, were also densely immunolabeled.

Due to its low activation threshold, it is strongly believed that Cav1.3 is one of the main channel types that mediate the persistent inward currents (PICs) in spinal motoneurons. Although it is still under debate as to where the PICs originate along the soma-dendritic membrane, physiological evidence suggests that these currents mainly originate from dendrites, although it can also originate from neuronal somata. In motoneurons the PICs could evoke all-or-none plateau potentials following injection of brief current pulses. With normal synaptic excitation there is evidence for a gradual recruitment of the PICs. Since the Cav1.3 channels in motoneurons are facilitated by monoaminergic innervation it has been suggested that the PICs mediated by Cav1.3 may serve as a controlled amplifier of synaptic excitation to motoneurons.

Key words: L-type calcium channel, alpha 1D subunit, immunohistochemistry, persistent inward current, antibody

ABBREVIATIONS

5-HT	Serotonin
Ca^{2+}-PIC	calcium persistent inward current
CNS	central nervous system
DHP	dihydropyridine
IHC	immunohistochemistry
IR	immunoreactivity
ISH	in situ hybridization
LTCC	L-type calcium channel
Na^+-PIC	persistent inward current
PIC	persistent inward current

INTRODUCTION

Voltage-gated calcium channels play an important role in many fundamental physiological processes such as muscle contraction, hormone secretion, neurotransmission, and gene expression in a wide range of cell types [Ertel et al., 2000; Catterall et al., 2005]. So far, five subtypes of calcium channels have been cloned, which include L, N, P/Q, R and T types. Among these, the L-type calcium channels (LTCCs) perhaps are the best characterized as essential for coupling excitation to contraction in skeletal, cardiac and smooth muscle cells [Schneider and Chandler, 1973; Reuter, 1985; Beam, et al., 1989; Tanabe et al., 1990; Franzini-Armstrong and Protasi, 1997]. Later it was found that LTCCs are also expressed in neurons and endocrine cells where they regulate many processes including hormone secretion and neurotransmitter release [Smith et al. 1993; Tachibana et al. 1993; Ashcroft et al. 1994; Dunlap et al. 1995; Charles et al. 1999; Wiser et al. 1999; Sand et al. 2001; Ouardouz et al. 2003; Jacobo et al., 2009]. Voltage-gated calcium channels consist of a principle α_1 subunit and several auxiliary subunits, β, $\alpha_2\delta$, and γ, which have regulatory functions (Fig. 1A). The α_1 subunit is the largest subunit and it incorporates the conduction pore, the voltage sensor and gating apparatus, and the known sites of channel regulation by second messengers, drugs and toxins [Catterall et al., 2005]. It includes four homologous domains (I-IV) with six transmembrane segments (S1-S6, Fig. 1A). So far four LTCC α_1 subunits have been cloned, which include Cav1.1, Cav1.2, Cav1.3, and Cav1.4 (formerly α_{1S}, α_{1C}, $\alpha_{1D,}$ and α_{1F}) [Ertel et al., 2000; Catterall et al., 2005]. While Cav1.1 and Cav1.4 are mainly expressed in skeletal muscles and the retina respectively, Cav1.2 and Cav1.3 are widely expressed throughout the central nervous system (CNS), sensory and endocrine cells, excitable cells in the cardiovascular system and cochlear hair cells [Reuter, 1985; Beam et al ., 1989; Tanabe et al., 1990; Hell et al., 1993; Westenbroek et al., 1998; Morgans, 1999; Yang et al., 1999; Platzer et al., 2000; Namkung et al., 2001; Brandt et al., 2003; Mangoni et al., 2003; Lipscombe et al., 2004; Hatano et al., 2006; Zhang et al., 2008a; Sukiasyan et al., 2009; Comunanza et al., 2010; Zhang et al., 2011].

Although Cav1.2 and Cav1.3 are usually expressed in the same tissues, their biophysical and pharmacological properties are distinct. For example, Cav1.3 channels are activated at relatively hyperpolarized membrane potentials, usually at -55 mV, which is about 20-25 mV more hyperpolarized than for activation of Cav1.2 channels (Fig. 1B) [Koschak et al., 2001; Safa et al., 2001; Scholze et al., 2001; Xu and Lipscombe, 2001]; Cav1.3 channels are less sensitive to dihydropyridine (DHP) antagonists (nifedipine, nimodipine) than Cav1.2 channels (Fig. 1C) [Koschak et al., 2001; Xu and Lipscombe, 2001; Lipscombe et al., 2004]; Cav1.3 channels open and close with rapid kinetics relative to Cav1.2 [Lipscombe et al., 2004; Helton et al., 2005]. All these properties suggest that Cav1.3 channels mediate subthreshold calcium signaling and thereby drive oscillatory activity in several excitable cells. For example, DHP antagonists suppress spontaneous intracellular calcium oscillations and slow rhythmic firing in Purkinje neurons, suprachiamatic neurons, pituitary cells, midbrain dopamine neurons and adrenal chromaffin cells [Charles et al., 1999; Giraldez et al., 2002; Liljelund et al., 2000; Pennartz et al., 2002; Placantonakis and Welsh, 2001; Vergara et al., 2003; Jackson et al., 2004; Comunanza et al., 2010; MarCantoni et al., 2010]. Studies using Cav1.3 knockout mice have partly confirmed these findings. For example, these mice always show sinoatrial node dysfunction and arrhythmia [Platzer et al., 2000; Zhang et al., 2002; Clark et al., 2003; Mangoni et al., 2003]. Also, due to their slow voltage-dependent inactivation Cav1.3 channels could mediate sustained calcium ion (Ca^{2+}) entry during the "non-inactivating" current (persistent inward current, PIC, seen as plateau potential in current clamp mode). This plateau property has been demonstrated to exist in spinal neurons in many vertebrates including turtles, mice, rats, cats and perhaps humans [Schwindt and Crill, 1977,

1980; Hounsgaard et al., 1988; Hounsgaard and Mintz, 1988; Eken and Kiehn, 1989; Hounsgaard and Kjaerulff, 1992; Morisset and Nagy, 1996, 1998; Russo and Hounsgaard, 1996; Russo et al., 1997; Carlin et al., 2000b; Bennett et al., 2001; Gorassini et al., 2002a,b; Hultborn et al., 2004; Smith and Perrier, 2006; Theiss et al., 2007]. To a large extent plateau potentials are mediated by LTCCs, in which Cav1.3 is the important one [Hounsgaard and Mintz, 1988; Svirskis and Hounsgaard, 1998; Carlin et al., 2000b; Alaburda et al., 2002; Simon et al., 2003].

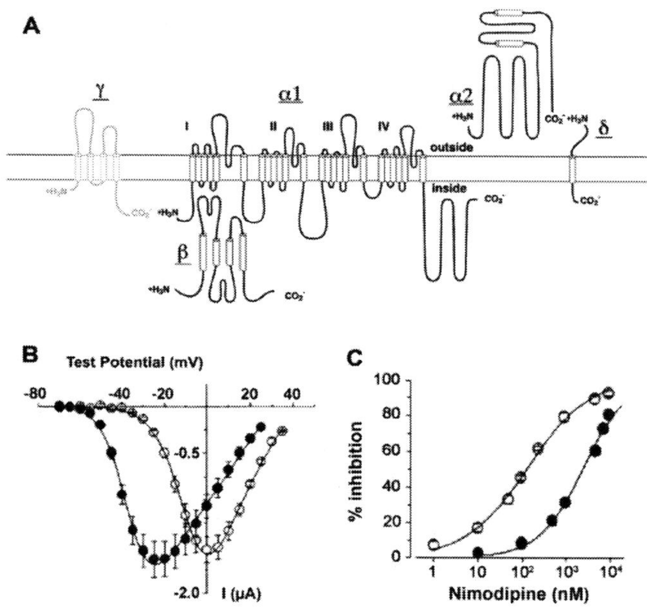

Figure 1. The structure of the subunits of voltage-gated calcium channels and some biophysical and pharmacological properties of two types of L-type calcium channels, Cav1.2 and Cav1.3. (A) The calcium channels consist of the principle α_1 subunit and several auxiliary subunits, β, $\alpha_2\delta$, and γ. The α_1 subunit consists of four homologous domains I-IV, each containing six transmembrane segments S1-S6 and a pore-forming region between segments S5 and S6. The Cav1.3 α_1 subunit is probably only regulated by β and $\alpha_2\delta$ but not γ subunits [Lacinova, 2005]. (B) Averaged peak current-voltage plots for Cav1.3α_1 (●) and Cav1.2α_1 (○) showing that they become activated at different voltages with 5 mM barium as charge carrier. Cav1.2α_1 currents begin to activate at approximately −35 mV whereas Cav1.3α_1 currents activate at approximately −55 mV. (C) Dose-response curve of nimodipine inhibition of Cav1.3α_1 (●) and Cav1.2α_1 (○) channel currents. It is apparent that Cav1.3α_1 is less sensitive to nimodipine than Cav1.2α_1. (A, adopted from Catterall et al., 2005, with copyright permission from Aspet; B and C, adopted from Xu and Lipscompe, 2001, with copyright permission from J Neurosci).

What then is the molecular mechanism by which LTCCs facilitate the plateau current? A number of intracellular messengers have been revealed to control both the pore-forming α_1 and the regulatory ($\alpha_2\delta$ and β for Cav1.2 and Cav1.3) subunits. The β subunit can have a facilitating effect on the PIC by shifting the activation curve to more hyperpolarized potentials. It also interacts with other proteins involved in second-messenger signaling pathways as well as cytoskeletal structures, causing the reorganization and incorporation of the target channels into the membrane and thus increasing the density of the channels conducting persistent Ca^{2+} current [Arikkath and Campbell, 2003; Dolphin, 2003; Catterall et al., 2005]. Calmodulin is known to be tethered constitutively to Cav1.2 and Cav1.3 channels, where it functions as a calcium detector responsible for both Ca^{2+}-dependent facilitation and inactivation [Halling et al., 2005; Perrier et al., 2000; Peterson et al., 1999]. Ca^{2+} persistent inward currents (Ca^{2+}-PICs) have also been shown to be promoted by stimulation of phosphatidylinositol 3-kinases [Viard et al., 2004] and protein kinase A pathways [Dai et al., 2009; Qu et al., 2005], whereas protein kinase C inhibits Cav1.3 [Baroudi et al., 2006].

To better understand the functional significance of Cav1.3 channels in a specific tissue it is essential to know their distribution in that tissue. Since the structure of the neuronal Cav1.3 α1 subunit was identified [Williams et al., 1992] the interest in exploring its expression in the CNS has increased. A number of studies have reported the distribution of Cav1.3 channels (mRNAs or proteins) in different parts of the brain, including the cerebellum, cerebral cortex, and hippocampus [e.g., Hell et al., 1993; Tanaka et al., 1995; Ludwig et al., 1997; Westenbroek et al., 1998; Takada et al., 2001; Grunnet and Kaufmann, 2004]. Several research groups including our own have investigated the expression of Cav1.3 in the spinal cords of many different species [Westenbroek et al., 1998; Jiang et al., 1999; Carlin et al., 2000b; Simon et al., 2003; Zhang et al., 2006, 2008a; Sukiasyan et al., 2009]. In addition, there are also a few simulation studies focusing on the distribution of Cav1.3 on motoneuron dendrites [Elbasiouny et al., 2005; Bui et al., 2006; Grande et al., 2007]. In this chapter we have systematically reviewed the distribution of Cav1.3 channels across different regions of the CNS based on the available published data and some unpublished data from our laboratory involving several different animal species. In addition we also discuss its potential functions with respect to motor control.

THE DISTRIBUTION OF THE CAV1.3 CHANNELS IN THE CNS

Due to the incompleteness of the data available from the literature on the distribution of Cav1.3 in some regions in the central nervous system we have decided to use some of our unpublished data in this chapter. These new data include: 1) Cav1.3 distribution pattern in the brain rostral to the midbrain including the cerebellum from two adult Wister rats. Several series of transverse sections were immunolabeled with a polyclonal Cav1.3 antibody from Sigma-Aldrich (see Table 1). 2) Cav1.3 immunolabeling pattern in spinal cords from 4 adult mice (C57BL/6J) and 2 adult turtles (*Chrysemys scripta elegans,* kindly provided by Dr. Jean-Francois Perrier, University of Copenhagen) using several different Cav1.3 antibodies to test whether the same antibody from different sources produces different labeling on tissue from the same species. 3) Cav1.3 immunolabeling on lumbar spinal motoneurons of cats intracellularly labeled with Neurobiotin. The immunolabeling methods were either avidin-biotin complex peroxidase or fluorescent immunohistochemistry (IHC) of which the detailed procedure has been described elsewhere [Zhang et al., 2006, 2008a; Sukiasyan et al., 2009]. All experimental procedures were conducted in accordance with the guidelines of *EU Directive 86/609/EEC* and were approved by the Danish Animal Experimentation Inspectorate. Because Cav1.3 expression in the spinal cord has been thoroughly studied in several different animal species we will first describe this channel expression across the spinal cord followed by the different brain regions. The nomenclature of the anatomical structures in the brain primarily refers to the rat brain atlas published recently by Paxinos and Watson (2007).

Table 1. Comparison of frequently used Cav1.3 antibodies[1]

Cav1.3 antibody source	Catalogue number (custom antibody name); host and clone	Antigen origin; peptide residue position	Immunolabeling characteristics	References
Custom-made (Univ. of Washington)	Anti-CND1; Rabbit polyclonal	Rat Cav1.3; 809 - 825	Mainly labels the cell somata and proximal dendrites both in the brain and spinal cord of rats.	Hell et al., 1993; Westenbroek et al., 1998.
Millipore-Chemicon (Temecula, CA, USA)	AB5158; Rabbit polyclonal	Rat Cav1.3; 809 – 825 and 859 - 875	Batch-dependent variations. The antibodies purchased before 2007 (e.g. lot: 25020248) mainly label the cell somata and proximal dendrites both in the brain and the spinal cord of rats and mice. Those purchased after 2007 (e.g. lot: JC1682903) label more distal dendrites.	Hanson and Smith, 2002; Anelli et al., 2007; Zhang et al., 2007; Leitch et al., 2009; Sukyasian et al., 2009.
Alomone Labs (Jerusalem, Israel)	ACC-005; Rabbit polyclonal	Rat Cav1.3; 809 – 825 and 859 - 875	Mainly labels the cell somata and proximal dendrites both in the brain and spinal cord of rats. On turtle spinal cord it seems mainly labels axonal terminals[2].	Jiang et al., 1999; Carlin et al., 2000b; Chung et al., 2000; Takada et al., 2001; Simon et al., 2003; Dobremez et al., 2005.
Custom-made (Univ. of Texas)	AM9742; Rabbit polyclonal	Rat Cav1.3; 2169 – 2203, from a long splice isoform	Mainly labels the cell somata and proximal dendrites both in the brain and spinal cord of cats and rats.	Zhang et al., 2006, 2008a,b.
Custom-made (Univ. of Copenhagen)	Not available; Rabbit polyclonal	Rat Cav1.3 2121 - 2137	Mainly labels the cell somata and proximal dendrites at least in the hippocampus and cerebellum of rats.	Grunnet and Kaufmann, 2004.
Sigma-Aldrich (St. Louis, MO, USA)	C1728; Rabbit polyclonal	Rat Cav1.3; 809 - 825	Mainly labels the cell somata and proximal dendrites both in the brain and spinal cord of rats and mice.	Unpublished data presented in this chapter.
Santa Cruz Biotechnology Inc (Santa Cruz, CA, USA)	sc-25687 Rabbit polyclonal	Human Cav1.3; 1661-1900	Labels cat, rat, mouse or turtle spinal neurons but with very high background level.	Unpublished data presented in this chapter.
Santa Cruz Biotechnology Inc (Santa Cruz, CA, USA)	sc-16251; Goat polyclonal	Human Cav1.3; Carboxy terminus	Does not label cat, rat or mouse spinal cord; labels motoneuron somata and dendrites of turtle spinal cord.	Unpublished data presented in this chapter.

1. This table contains summarized data based on past research and the authors are not responsible for diverse results if taken as a guideline in future research.
2. Cav1.3 antibodies from Alomone Labs, Millipore-Chemicon and Sigma-Aldrich produced a similar immunolabeling pattern on turtle spinal cord.

1.　Spinal cord

The expression of Cav1.3 in the spinal cord has been reported in several animal species including cat, rat, mouse and turtle [Westenbroek et al., 1998; Jiang et al., 1999; Carlin et al., 2000b; Simon et al., 2003; Dobremez et al., 2005; Zhang et al., 2006, 2008a; Anelli et al., 2007; Sukiasyan et al., 2009]. For the cat and rat thorough studies of Cav1.3 expression have been performed using the IHC technique by us [Zhang et al., 2006, 2008a; Sukiasyan et al., 2009]. For the mouse and turtle the expression of Cav1.3 had only been investigated in the ventral horn motoneuron region so far [Jiang et al., 1999; Carlin et al., 2000b; Simon et al., 2003]. From the published data it appeared that expression differences exist between the different species with respect to the channel localization and the spatial distribution along the soma-dendritic axes in spinal motoneurons. However, due to the different antibodies used and a possible problem of antibody specificity in different animal tissues (see section below) it is hard to make conclusions on species differences from a direct comparison of Cav1.3 immunolabeling using different antibodies. Below we will describe the Cav1.3 channel expression in spinal cords in different animals (cat, rat, mouse, and turtle) based on published data and, in the cases where the data were not sufficient, unpublished data from our laboratory.

1)　Cat

We have investigated distribution of Cav1.3-immunoreactivity (IR) across the cat spinal cord using a custom-made polyclonal antibody that recognizes the long splice variant of rat Cav1.3 subunit (Cav1.3a, Table 1) [Zhang et al., 2006, 2008a]. The results showed that Cav1.3-IR was widely distributed in all segments of the spinal cord but the frequency of immunolabeled neurons in the different laminae of the spinal gray matter varied, with the highest frequency of the immunolabeled neurons being in the ventral horn motoneuron region (laminae XIII and IX, ~100% of its neurons were Cav1.3 immunoreactive) and the lowest in the superficial dorsal horn (laminae I-III, ~7% of its neurons were Cav1.3 immunoreactive; Fig. 2). The immunolabeling intensity was highest in neuronal somata, but a certain length of the proximal dendrites was also immunolabeled. The neurons in the lateral ventral horn exhibited a particularly intense immunolabeling, which include the motoneurons innervating the limb muscles as well as some specified motoneuronal groups such as the phrenic nucleus in the cervical segments and the Onuf's nucleus in the lower lumbar and upper sacral segments.

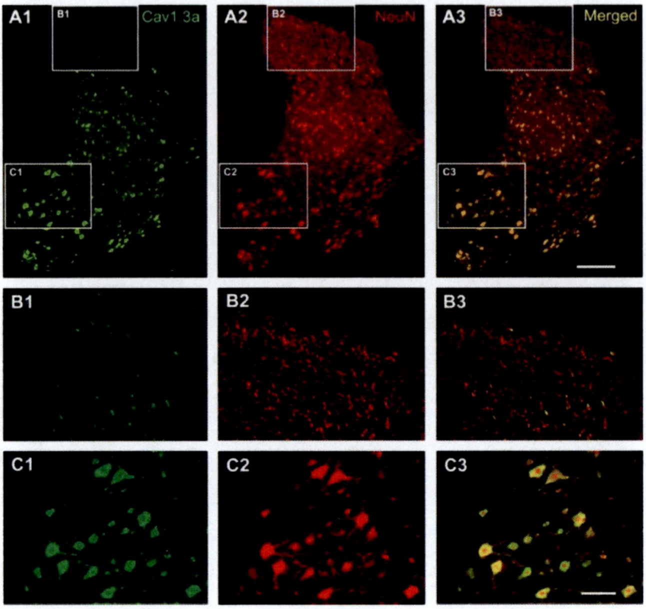

Figure 2. Epifluorescent photomicrographs from a NeuN (red) and Cav1.3a (green) double-labeled cervical section showing details of Cav1.3 immunoreactivity across different laminae in a cat spinal section (**A1–A3**). The two lower rows are enlargements of the regions from the upper row demarcated by rectangles showing the detailed neuronal labeling in the superficial dorsal horn (**B1–B3**) and the lateral ventral horn (**C1–C3**). Note that only a small portion of neurons in the superficial dorsal horn (B3) appear to be double labeled (yellow), whereas all neurons in the lateral ventral horn (C3) are double labeled. Scale bars in A3 (for A1–A3): 400 µm; in C3 (for B1–C3): 150 µm. (Adopted from Zhang et al., 2008a).

2) Rat

Westenbroek et al. [1998] were the first to investigate the expression of Cav1.3 across selected rat spinal segments using a custom-made Cav1.3 antibody (Table 1). They showed that Cav1.3 was expressed in neurons both in the dorsal and ventral horn. A further detailed description about its distribution in different segments and gray matter laminae was not provided. Dobremez et al. [2005] only reported Cav1.3-IR in the rat lumbar dorsal horn in relation to Cav1.3 expression changes after sciatic nerve injury. Anelli et al. [2007] reported Cav1.3 expression changes in the rat sacral motor neurons following spinal cord transection. We have made a systematic investigation of Cav1.3-IR using a commercially available Cav1.3 antibody (Table 1) and demonstrated that Cav1.3-immunoreactive neurons were widely distributed in all parts of the spinal cord (Fig. 3) [Sukiasyan et al., 2009]. In the spinal grey matter the intensity of Cav1.3-IR varied noticeably for neurons in different locations. In contrast to the immunolabeling pattern in the cat spinal cord where no glial cells were detected to contain Cav1.3-IR, in the rat spinal cord a large proportion of oligodendrocytes were immunoreactive to the Cav1.3 antibody. Just as is seen in cat tissue, the large neurons in the ventral horn (presumably motoneurons) of the rat spinal cord tended to display higher levels of Cav1.3-IR whereas the small neurons in the dorsal horn displayed lower levels.

Whilst the Cav1.3-IR was most dense in the neuronal somata, the first and second order dendrites of large (moto)neurons were also clearly immunolabeled for Cav1.3.

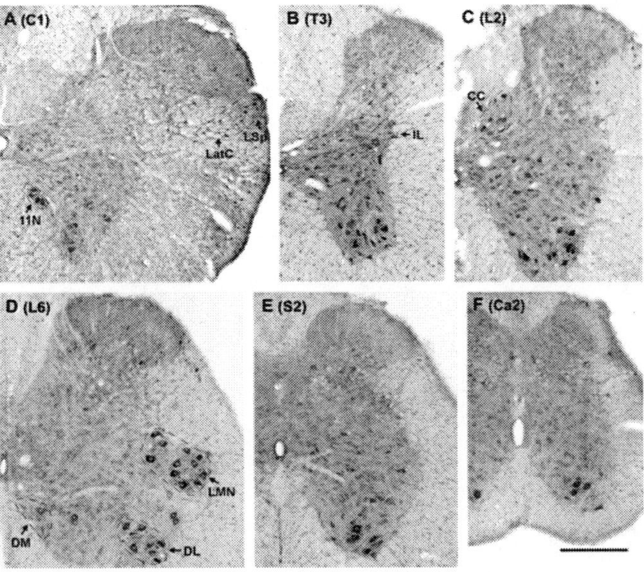

Figure 3. Low-power microphotographs of Cav1.3 immunoreactivity in selected segments of the rat spinal cord. (A) Cervical segment; (B) Thoracic segment; (C, D) Lumbar segments; (E) Sacral segment; (F) Caudal segment. The segmental level is indicated in parenthesis. One common characteristic is that the large neurons in the ventral horn always showed more intense labeling than the small neurons in other regions. These large neurons in some segments, e.g., lower lumbar segment (D), usually form specialized neuronal clusters. Several specialized neuronal groups are indicated with dash lines in A-D. Abbreviations: 11N: nucleus of accessory nerve; CC: Clarke's column; DL: Dorsolateral part of the Onuf's nucleus. DM: Dorsomedial part of the Onuf's nucleus; IL: intermediolateral nucleus; LatC: lateral cervical nucleus; LMN: lateral motoneuron group; LSp: lateral spinal nucleus; Scale bar in F: 400 μm.

3) Mouse

The first study showing the expression of the Cav1.3 channel and its changes in expression during development in the mouse spinal cord came from Brownstone's laboratory [Jiang et al., 1999]. Using a Cav1.3 antibody from Alomone Labs (Table 1) they showed that in the (lumbar) spinal cord Cav1.3 began to be expressed in the nuclei of some cells at postnatal day 3. Thereafter the expression increased with age and approached an adult pattern by postnatal day 18, where Cav1.3 was predominantly expressed in the distal (second and third order) dendrites [Carlin et al., 2000b]. In our laboratory we have just performed a Cav1.3 IHC study on mouse spinal cord similar to that which we have performed in the cat and rat to see whether there are any expression differences between the different animal species. We used rabbit Cav1.3 polyclonal antibodies from Millipore-Chemicon from several different batches (purchased from different times; see Table 1). The results showed a widespread distribution of Cav1.3-IR in different segments of the mouse spinal cord. A large portion of neurons in the spinal gray matter were Cav1.3 immunoreactive and the larger neurons in the ventral horn (probably motoneurons) displayed a higher intensity of immunolabeling similar to the rat.

Nevertheless, the results were somewhat confusing. By using the antibody from the same batch as used in the rat study (purchased in year 2005) a similar labeling pattern was seen in the mouse as in the rat with immunolabeling in the neuronal somata and dendrites (Fig. 4A), whereas by using the same antibody purchased recently (after year 2007, with a different lot number) Cav1.3-IR was predominantly detected in the neuronal dendrites (including both proximal and distal parts) although it could also be detected in the neuronal somata (Fig. 4B). However, when compared with the results from using the earlier antibodies the somatic immunolabeling intensity was much lower. It will be discussed below as to how the different labeling patterns could possibly be related to different antibodies.

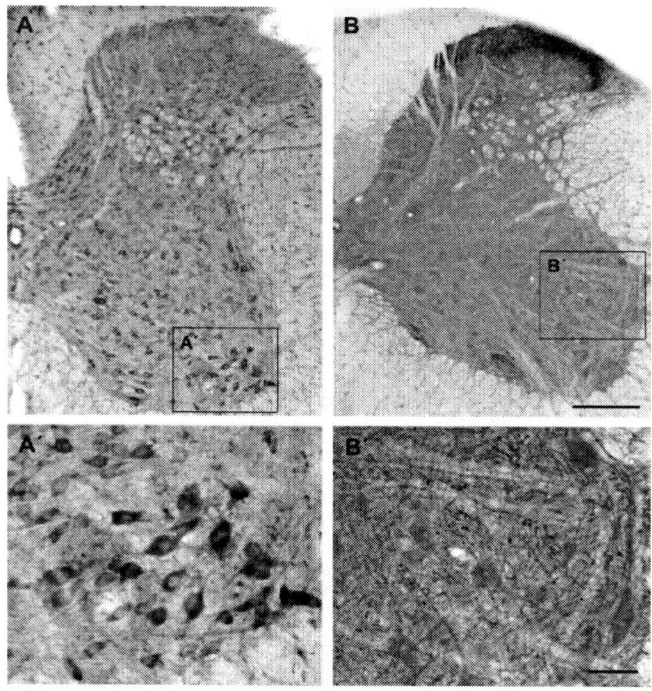

Figure 4. Cav1.3 immunoreactivity in a lumbar spinal section of the mouse spinal cord immunostained with antibodies from Millipore-Chemicon from two different batches. One antibody (old batch, lot #: 25020248, purchased in 2005) mainly immunolabeled the neuronal somata and their proximal dendrites throughout the gray matter (A, A′). The other antibody (new batch, lot #: JC1682903, purchased in 2009) densely immunolabeled the dendrites of different sizes in different parts of the gray matter although the neuronal somata were also immunolabeled but with a lower density when compared with those immunolabeled using the first antibody (B, B′). (A′) and (B′) are the enlargements of the areas demarcated in (A) and (B) respectively. Scale bar in B (for A and B): 200 μm; in B′ (for A′ and B′): 50 μm.

4) Turtle

By electrophysiological techniques Hounsgaard and coworkers have suggested that PICs in turtle motoneurons are mediated by Cav1.3 channels [Hounsgaard and Mintz, 1988; Alaburda et al., 2002; Perrier et al., 2002] and that the dominant part of this current is generated in the distal dendrites [Hounsgaard and Kiehn, 1993; Delgado-Lezama et al., 1999; Svirskis et al

2001]. Later using an antibody from Alomone Labs they demonstrated the expression of Cav1.3 channels in turtle (*Chrysemys scripta elegans*) lumbar spinal motoneurons [Simon at el., 2003]. In that study they have shown patches of Cav1.3 immunoreactive products in association with the membrane of choline acetyltransferase-labeled motoneuronal somata and both proximal and distal dendrites. Although this channel distribution pattern along the soma-dendritic surface fits with the physiological findings, the detailed localization of the channels was somewhat intriguing. By careful inspection of Figs. 3 and 5 in this article it is obvious that the Cav1.3-IR seems to be located completely on the surface of the cell somata and dendrites whereas in the cytoplasm Cav1.3-IR was lacking. In other species, e.g., in the cat the Cav1.3-IR has been detected abundantly in the cytoplasm [Zhang et al., 2008a]. A later study on the turtle spinal cord using electron microscopy from the same laboratory has found that almost all of the Cav1.3-IR was located at the presynaptic terminals [cited in Zhang et al., 2008a]. From where these Cav1.3-positive terminals come from is unknown. In any way, it is hard to explain the terminal localization of Cav1.3 in relation to the generation of PICs in the postsynaptic motoneurons. One probable explanation for this immunolabeling pattern in the turtle spinal cord is that the antibody used was not specific to the turtle Cav1.3 channel protein, the sequence of which has not been cloned (searched on The Protein Data Bank). Thus it is necessary to test several different antibodies and to do strict specificity controls for each of these antibodies. For adding some useful data to this chapter we have tried several commercial Cav1.3 antibodies from different companies (see Table 1 for detailed information) on turtle spinal cord and found that Cav1.3 antibodies from Alomone Labs, Millipore-Chemicon, and Sigma-Aldrich produced the same immunolabeling pattern as presented by Simon et al. [2003] whereas two different antibodies from Santa Cruz Biotechnology Inc apparently immunolabeled the neuronal somata and dendrites although they also immunolabeled somewhat a different pattern (Fig. 5). Because we did not perform specificity controls for these antibodies on the turtle spinal tissue we cannot claim which antibody reflected the real localization of Cav1.3 channels in turtle spinal cord.

2. Brain

In 1990 researchers from Catterall's laboratory [Ahlijanian et al., 1990; Westenbroek et al., 1990] first revealed the distribution and subcellular localization of DHP-sensitive L-type channels in the rat brain using a custom-made mouse monoclonal antibody (MANC1) that recognizes the $\alpha_2\delta$ subunits of the skeletal muscle calcium channel. They found that MANC1-positive neurons were expressed in many different brain regions including the cerebral cortex, hippocampus, olfactory bulb, and cerebellum. However, because this antibody was not specific for Cav1.3 it is hard to judge whether the labeling pattern was from Cav1.2, Cav1.3 or both. Later Hell et al. [1993] from the same group produced two polyclonal antibodies (anti-CNC1 and anti-CND1) specifically targeting brain Cav1.2 and Cav1.3, respectively. Using the immunoprecipitation technique they found that Cav1.2 accounted for 75% of the LTCCs in the cerebral cortex and hippocampus in rats and Cav1.3 accounted for only 20%.

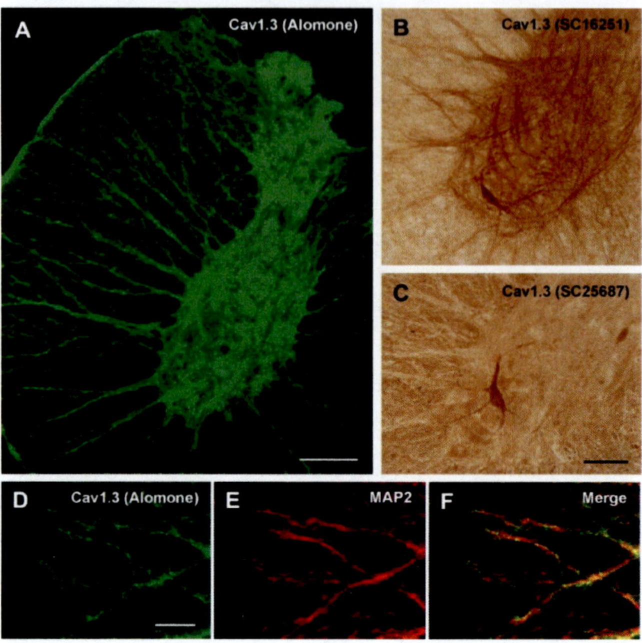

Figure 5. Cav1.3 immunoreactivity in the lumbar spinal cord of the turtle immunostained with three different Cav1.3 antibodies from different sources. (A) A spinal section immunostained with a Cav1.3 antibody from Alomone Labs showing that Cav1.3 immunoreactivity was widely distributed in the spinal cord, including the gray and white matter. The Cav1.3 immunoreactive profiles were consisted of beaded terminal-like components. No neuronal somata were observed to be labeled. (D) Higher power photomicrograph from another section showing the detailed Cav1.3 immunolabeling (green) in the lateral funiculus. (E) MAP2 (Microtubule-associated protein 2, a dendritic marker, red) double immunostaining on the same section. (F) Overlay of D and E showing that the Cav1.3-immunoreactive beads were spread along the surface of the dendrites but they were apparently not the same structures (no colocalization). (B, C) Immunolabeling patterns with two different Cav1.3 antibodies from Santa Cruz Biotechnology Inc showing that these antibodies labeled neuronal somata and dendrites in the ventral horn although the labeling patterns were also not the same. Scale bar in A: 200 µm; in C (for B and C): 100 µm; in D (for D-F): 50 µm.

Using an anti-CND1 antibody they found that Cav1.3 was expressed throughout different brain regions. However, in this study they only described the detailed labeling pattern in the cerebral cortex (dorsal cortex in their paper), hippocampus, and cerebellum, but not other brain regions. Thereafter several research groups have examined Cav1.3 expression pattern using either IHC or in situ hybridization (ISH) in different brain regions in rats [Tanaka et al., 1995; Ludwig et al., 1997; Chung et al., 2000; Takada et al., 2001; Hanson and Smith 2002; Grunnet and Kaufmann, 2004; Leitch et al., 2009; Sukiasyan et al., 2009]. However, a thorough investigation of Cav1.3 expression in the entire brain is still lacking. Hetzenauer et al. [2006] have examined Fos expression pattern in the brain elicited by the LTCC activator BayK8644 in wild type and $Cav1.2^{DHP-/-}$ mice. They found that following BayK8644 stimulation, Fos expression in wild type mice was significantly increased in about 96% (77/80) of examined brain regions, whereas in $Cav1.2^{DHP-/-}$ mice the value was only 25% (20/80). However, it is difficult to judge whether these functional expression data reflected the physical anatomical distribution of the channels in the brain. Thus, from the available data it can be concluded that although there are quite a few studies to investigate Cav1.3 localization in the brain, the description of the channel distribution in different brain regions is far from complete.

1) Olfactory bulb

Both ISH and IHC studies have detected a strong expression of LTCCs in the olfactory bulb in rats [Tanaka et al., 1995; Ludwig et al., 1997; Grunnet and Kaufmann, 2004]. However, a detailed expression profile was not provided in the IHC study as to which kind of cells in the olfactory bulb was immunoreactive for Cav1.3 [Grunnet and Kaufmann, 2004]. From the ISH studies it seemed that Cav1.3 was intensely expressed in the internal granule and mitral cells, whereas it was weakly expressed in the periglomerular cells [Tanaka et al., 1995; Ludwig et al., 1997].

2) Cerebral cortex (including hippocampus)

The available IHC and ISH data suggest that LTCCs (including Cav1.3) are widely distributed in the rat [Hell et al., 1993; Ludwig et al., 1997; Grunnet and Kaufmann, 2004] as well as in the mouse cerebral cortex [Timmermann et al., 2002] . However a detailed regional description regarding its distribution is lacking. In the cerebral cortex the pattern of Cav1.3 immunolabeling was only partially reported by Hell et al. [1993]. They described that in all regions of the dorsal cerebral cortex (without differentiating between specific regions) the Cav1.3 positive neurons were observed throughout layers I-VI. Their results also indicated that Cav1.3 was expressed with a relative high density in the cell somata and the proximal dendrites and with a low density in the distal dendrites. Our own IHC results in the rat cerebral cortex were in agreement with theirs. However, we should state that Cav1.3 positive neurons were throughout layers II-VI of the cerebral cortex because there were no cells being immunolabeled in the molecular layer (layer I; Fig. 6). This could also be discerned in Fig. 4 from the paper of Hell et al. [1993]. When examining the cerebral cortex in different regions it seemed that there were not apparent regional differences for the Cav1.3 expression, i.e., a similar immunolabeling pattern was seen in most of the cortical regions from the most occipital to most frontal cortices. However in different cell layers the densities of Cav1.3-IR were not the same. Cells in layers II-IV were usually more densely immunolabeled than the cells in layers V and VI (Fig. 6).

In the hippocampus Cav1.3 was expressed throughout the different areas and in several different types of neurons in rats [Hell et al., 1993; Grunnet and Kaufmann, 2004; Leitch et al., 2009], which include the pyramidal neurons in the CA1-CA3 areas and the interneurons in the strata oriens, radiatum, lacunosum moleculare, and dentate gyrus. The subcellular distribution pattern of Cav1.3-IR was similar to that observed in neurons in the cerebral cortex, i.e., a dense Cav1.3-IR was seen at the neuronal somata and the basal and apical dendrites, and a weak Cav1.3-IR was seen at the distal parts of the dendrites. However at the ultrastructural level, using the immunogold labeling technique, Cav1.3 immunolabeling has been revealed throughout the dendritic processes of all pyramidal neurons in the hippocampus [Leitch et al., 2009].

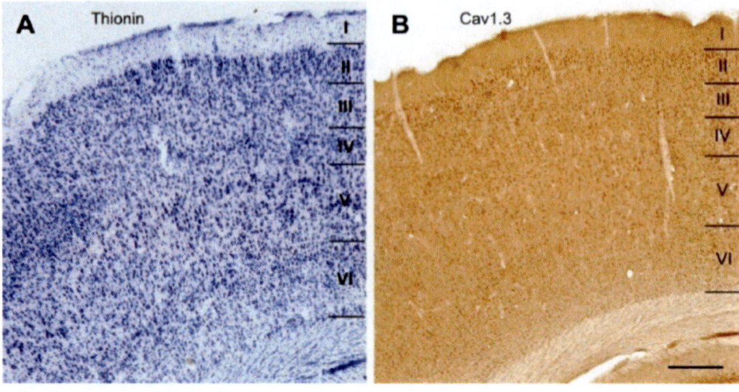

Figure 6. Transverse sections showing the Cav1.3 immunoreactivity in the frontal cerebral cortex. (A) Cav1.3 immunoreactive cells can be seen throughout layers II to VI but not in layer I. The density of immunolabeling density is higher in the superficial layers (II-IV) than in the deep layers (V and VI). (B) An adjacent section stained with thionin to show the cytoarchitecture of the cerebral cortex. Scale bar in B: 300 µm.

3) Basal ganglia

The anatomical studies of Cav1.3 expression in the basal ganglia are very scarce [Tanaka et al., 1995; Hanson and Smith, 2002] although several investigations have suggested the possible existence of the channel in these regions in rats [Surmeier et al., 1994; Grunnet and Kaufmann, 2004; Olson et al., 2005; Zhang et al., 2005; Hetzenauer et al., 2006]. Hanson and Smith [2002] revealed the expression of Cav1.3 channels in rat globus pallidus neurons and by electron microscopy they showed that the channel proteins were distributed in dendrites of all diameters. Zhang et al. [2005] using the technique of PCR showed that in a culture from rat caudate nucleus most of the medium spiny cells expressed Cav1.3. From our unpublished data from rats it can be seen that almost all of the cells (mostly medium spiny cells) in the accumbens nucleus, caudate nucleus and putamen were Cav1.3 immunoreactive (Fig. 7). The density of immunolabeling was moderate. In the globus pallidus most of the cells were also labeled but the density was lower in comparison to that observed in other striatal nuclei. Most cells in different parts of the amygdaloid complex, e.g., the central and lateral amygdaloid nuclei, were moderately immunolabeled (data not shown).

4) Diencephalon

Reports of Cav1.3 expression in the diencephalon are sporadic [Tanaka et al., 1995; Budde et al., 1998; Joux et al., 2001]. Using ISH Ludwig et al. [1997] did not find any Cav1.3 expression signals above the background level in the rat thalamus; whereas in the hypothalamus, the labeling signals, though above the background level, were low. Using IHC technique Grunnet and Kaufmann [2004] demonstrated that prominent immunolabeling was found in the rat thalamus. However, due to the low magnification of the microphotographs provided in the paper it is hard to identify which nuclei contained Cav1.3 immunoreactive cells. Joux et al. [2001] reported the expression of many types of calcium channels in the neurons of rat supraoptic nucleus, which include Cav1.2, Cav1.3, Cav2.1, Cav2.2, and Cav2.3 subunits. Budde et al. [1998] using a fluorescent DHP ratio technique found that LTCCs were widely distributed in thalamocortical relay neurons, interneurons and reticular neurons in rat thalamus. They also found that in thalamocortical relay neurons, LTCCs were clustered in high density around the base of dendrites, whereas they were more evenly distributed on the soma of interneurons. The neurons in the reticular nucleus exhibited a high density of L-type channels in central regions. However the authors did not specify what kind(s) of LTCC (Cav1.2, Cav1.3 or both) they have revealed.

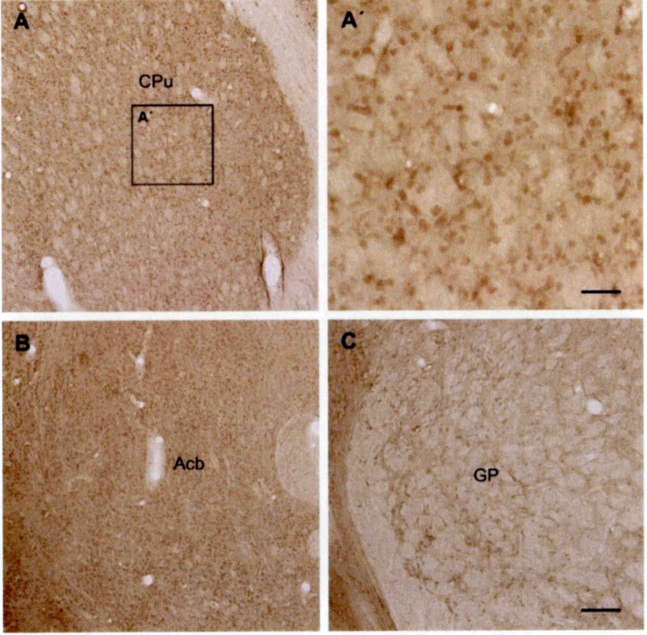

Figure 7. Transverse sections showing the Cav1.3 immunoreactivity in the striatum. (A) In the caudate-putamen. (A′) is an enlargement of the squire area in (A). (B) In the accumbens. (C) In the globus pallidus. It is obvious that all or at least most of the striatal cells (most are medium spiny neurons) were immunoreactive for Cav1.3. Scale bar in C (for A – C): 200 μm; in A′: 50 μm.

As stated above we have performed an IHC study in different brain regions in two rats. Here we will briefly report our immunolabeling results in this region. As shown in Fig. 8, Cav1.3 was widely expressed in the diencephalon. In the thalamus strongly immunolabeled nuclei included the medial and lateral habenular nuclei, reticular nucleus, some sensory relay nuclei (medial and lateral geniculate nuclei), the intralaminar nuclei (centrolateral, paracentral, and central medial nuclei), and the motor-related nuclei (ventral lateral and ventral medial nuclei). Some other nuclei, such as the ventral posterior medial and lateral nuclei, posterior nuclear complex, dorsal medial, ventral medial, anterior ventral, and anterior dorsal nuclei were moderately to weakly immunolabeled. Among all the nuclei in the thalamus, the medial habenula was the most strongly labeled nucleus. In the hypothalamus many nuclei were strongly labeled, such as the arcuate hypothalamic, dorsomedial hypothalamic, paraventricular hypothalamic, suprachiasmatic, and preoptic nuclei. Some other nuclei or regions were moderate to weakly labeled, such as the lateral hypothalamic and the anterior hypothalamic areas. In the subthalamic region the cells in the subthalamic nucleus were strongly immunolabeled and those in the zona incerta were moderately immunolabeled. In the diencephalon Cav1.3-IR was located mainly in the cell somata although a certain length of their dendrites, especially proximal dendrites, was also frequently immunolabeled.

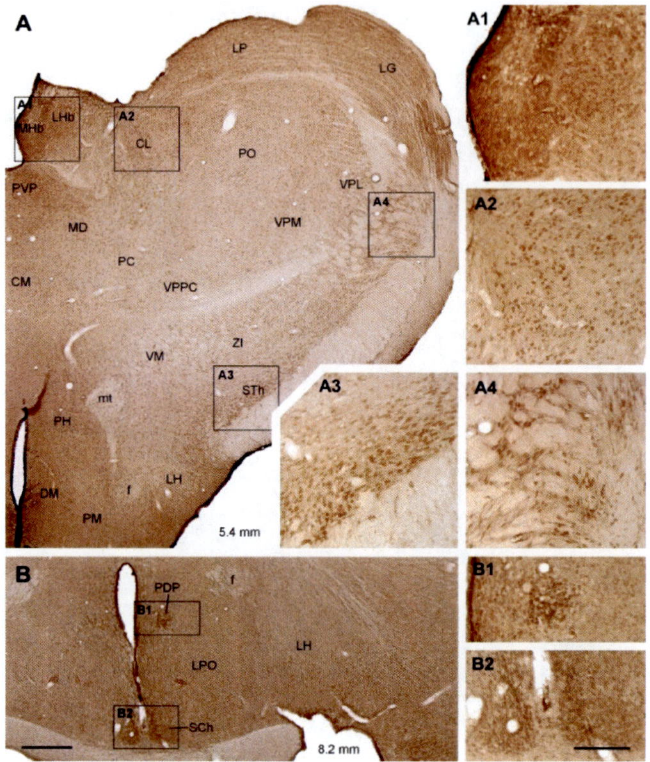

Figure 8. Transverse sections showing the Cav1.3 immunoreactivity in the diencephalon. (A, B) Low-power photomicrographs showing the Cav1.3 immunolabeling in the thalamus and hypothalamus. The Cav1.3 immunoreactive neurons were widely distributed in different parts of these regions but with differential labeling intensities. (A1-A4, B1, B2) Enlargement of the areas demarcated in (A) and (B) showing the detailed Cav1.3 immunolabeling pattern in these different regions (nuclei). Abbreviations: CL, centrolateral thalamic nucleus; CM, central medial thalamic nucleus; DM, dorsomedial hypothalamic nucleus; f, fornix; LG, lateral geniculate nucleus; LH, lateral hypothalamic area; LHb, lateral habenular nucleus; LP, lateral posterior thalamic nucleus; LPO, lateral preoptic area; MD, mediodorsal thalamic nucleus; MHb, medial habenular nucleus; mt, mammillothalamic tract; PC, paracentral thalamic nucleus; PDP, posterodorsal preoptic nucleus; PH, posterior hypothalamic area; PM, premammillary nucleus; Po, posterior thalamic nuclear group; PVP, paraventricular thalamic nucleus, posterior part; SCh, suprachiasmatic nucleus; STh, subthalamic nucleus; VM, ventromedial thalamic nucleus; VPL, ventral posterolateral thalamic nucleus; VPM, ventral posteromedial thalamic nucleus; VPPC, ventral posterior thalamic nucleus, parvicellular part; ZI, zona incerta. The numbers at the bottom-right corner of panels A and B indicate the section position relative to the interaural plane. Scale bar in A (for A and B): 500 μm; in B2 (for A1-A4, B1 and B2): 200 μm.

5) Brain stem

Although some studies have revealed the expression of Cav1.3 in the brain stem using ISH and IHC [e.g., Tanaka et al., 1995; Ludvig et al., 1997; Takada et al., 2001; Grunnet and Kaufmann, 2004] we were the only one who have described a detailed Cav1.3-IR distribution throughout different nuclei in this region in rats [Sukiasyan et al., 2009]. Below we will summarize our finding based on this work.

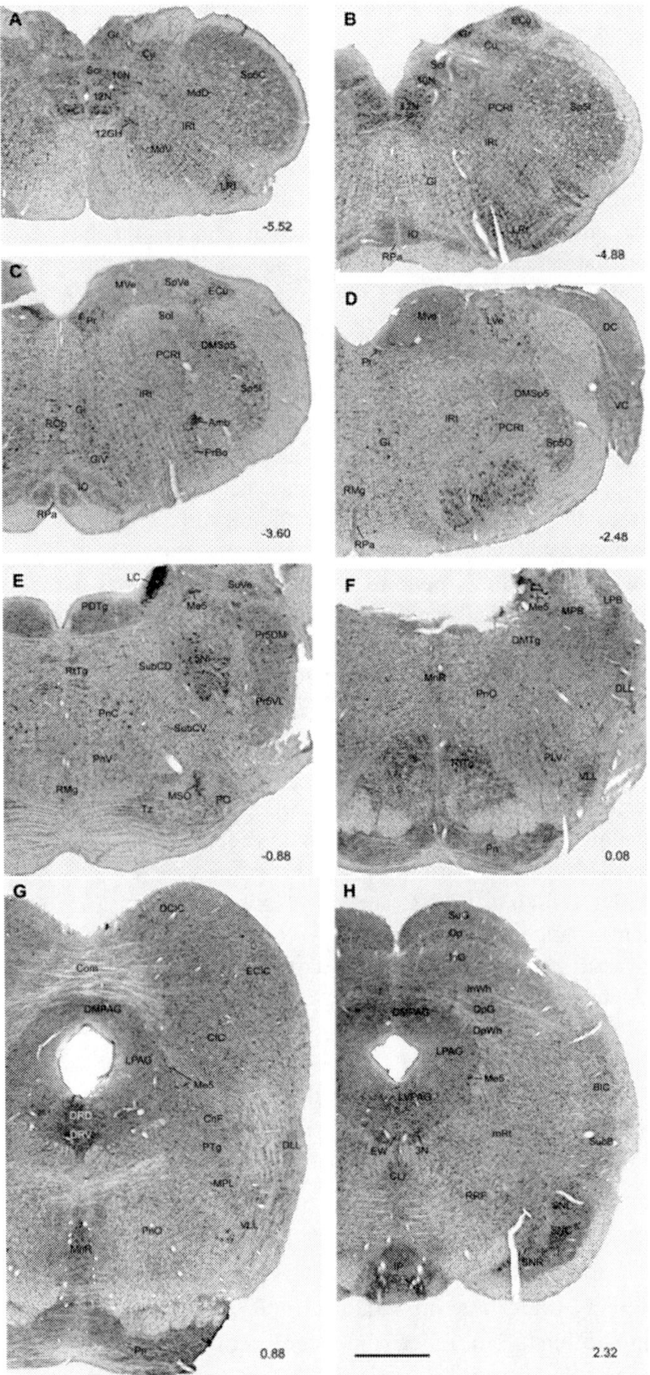

Figure 9. Low-power photomicrographs showing the Cav1.3 immunoreactivity in selected transverse sections from different rostrocaudal levels of the brain stem. (A, B, C) From the medulla oblongata. (D, E, F) From the pons. (G, H) From the midbrain. Note that the Cav1.3 immunoreactive neurons were

widely distributed across different parts of the brain stem but with noticeably different labeling intensities. Abbreviations: 3N, oculomotor nucleus; 5N, motor trigeminal nucleus; 7N, fasial nucleus; 10N, dorsal motor nucleus of vagus; 12GH, hypoglossal nucleus, geniohyoid part; 12N, hypoglossal nucleus; Amb, ambiguous nucleus; BIC, nucleus of the brachium of the inferior colliculus; CIC, central nucleus of the inferior colliculus; CLi, caudal linear nucleus of the raphe; CnF, cuneiform nucleus; Com, commissural nucleus of the inferior colliculus; Cu, cuneate nucleus; DC, dorsal cochlear nucleus; DCIC, dorsal cortex of the inferior colliculus; DLL, dorsal nucleus of the lateral lemniscus; DMPAG, dorsomedial periaqueductal gray; DMSp5, dorsomedial spinal trigeminal nucleus; DMTg, dorsomedial tegmental area; DpG, deep gray layer of the superior colliculus; DpWh, deep white layer of the superior colliculus; DRD, dorsal raphe nucleus, dorsal part; DRV, dorsal raphe nucleus, ventral part; ECIC, external cortex of the inferior colliculus; ECu, external cuneate nucleus; EW, Edinger-Westphal nucleus; Gi, gigantocellular reticular nucleus; GiV, gigantocellular reticular nucleus, ventral part; Gr, gracile nucleus; InG, intrmediate gray layer of the superior colliculus; InWh, intermediate white layer of the superior colliculus; IO, inferior olive; IP, interpeduncular nucleus; IRt, intermediate reticular nucleus; LC, locus coeruleus; LPAG, lateral periaqueductal gray; LPB, lateral parabrachial nucleus; LRt, lateral reticular nucleus; LVe, lateral vestibular nucleus; LVPAG, lateroventral periaqueductal gray; MdD, medullary reticular nucleus, dorsal part; MdV, medullary reticular nucleus, ventral part; Me5, mesencephalic trigeminal nucleus; MnR, median raphe nucleus; MPB, medial parabrachial nucleus; MPL, medial paralemniscial nucleus; mRt, mesencephalic reticular formation; MSO, medial superior olive; MVe, medial vestibular nucleus; Op, optic nerve layer of the superior colliculus; PCRt, parvicellular reticular nucleus; PDTg, posterodorsal tegmental nucleus; PLV, perilemniscal nucleus, ventral part; Pn, pontine nuclei; PnC, pontine reticular nucleur, caudal part; PnO, pontine reticular nucleus, oral part; PnV, pontine reticular nucleus, ventral part; PO, periolivary nucleus; Pr, prepositus nucleus; Pr5DM, principal sensory trigeminal nucleus, dorsomedial part; Pr5VL, principal sensory trigeminal nucleus, venrolateral part; PrBo, pre-Bötzinger complex; PTg, pedunculopontine tegmental nucleus; RMg, raphe magnus nucleus; Rob, raphe obscurus nucleus; RPa, raphe pallidus nucleus; RRF, retrorubral field; RtTg, reticulotegmental nucleus of the pons; SNC, substansia nigra, compact part; SNL, substansia nigra, lateral part; SNR, substansia nigra, reticular part; Sol, nucleus of the solitary tract; Sp5C, spinal trigeminal nucleus, caudal part; Sp5I, spinal trigeminal nucleus, interpolar part; Sp5O, spinal trigeminal nucleus, oral part; SpVe, spinal vestibular nucleus; SubB, subbrachial nucleus; SubCD, subcoeruleus nucleus, dorsal part; SubCV, subcoeruleus nucleus, ventral part; SuG, superficial gray layer of the superior colliculus; SuVe, superior vestibular nucleus; Tz, nucleus of the trapezoid body; VC, ventral cochlear nucleus; VLL, ventral nucleus of the lateral lemniscus. The number at the bottom-right corner of each panel indicates the section position relative to the interaural plane. Scale bar in H: 1.0 mm. (Adopted from Sukiasyan et al., 2009).

Cav1.3-IR was widely distributed in many different nuclei or structures in the brain stem. However the proportion (frequency) of Cav1.3 positive neurons and the intensity of Cav1.3-IR varied considerably between different nuclei (Fig. 9). In some nuclei all or almost all of the neurons were immunolabeled, e.g., most of the cranial motor nuclei; but in some other nuclei only a portion of their neurons were immunolabeled, e.g. in some sensory nuclei and nuclei in the reticular structure. Usually the neurons in motor-related nuclei displayed higher immunolabeling intensity than those in sensory nuclei. The cranial motor nuclei that showed high immunolabeling intensity included the occulomotor, trochlear, and Edinger-Westphal nuclei in the midbrain; the motor trigeminal, abducens, and facial nuclei in the pons; the dorsal motor nucleus of vagus, the nucleus of the accessory nerve, and the hypoglossal nucleus in the medulla. The sensory nuclei that showed a moderate intensity of immunolabeling included different trigeminal sensory nuclei in the pons and medulla (the principle trigeminal and spinal trigeminal nuclei), the vestibular and cochlear nuclear complexes in the pons. The mesencephalic trigeminal nucleus in the midbrain was more intensely immunolabeled than the other sensory nuclei. Many other nuclei including many reticular nuclei in different brain stem regions (e.g., gigantocellular reticular nucleus, lateral reticular nucleus, and reticulotegmental nucleus of the pons), inferior olive, pontine nucleus, red nucleus, and periaqueductal gray were moderately to intensely immunolabeled.

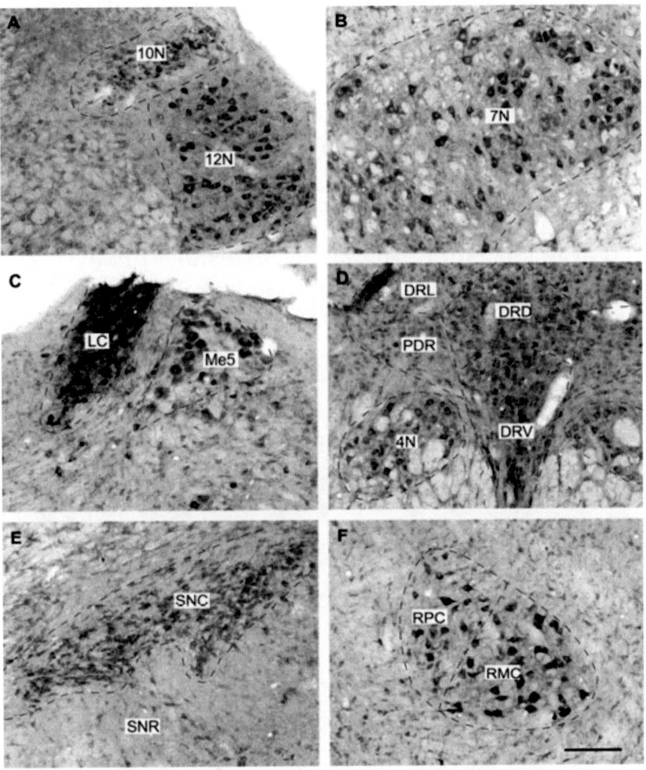

Figure 10. High-power photomicrographs showing the Ca$_V$1.3 immunoreactivity in some selected nuclei of interest (outlined with dash lines) in the brain stem. (A) The dorsal motor nucleus of vagus (10N) and the hypoglossal nucleus (12N). (B) The facial nucleus (7N). (C) The locus coeruleus (LC) and the mesencephalic trigeminal nucleus (Me5). (D) The trochlear (4N) and the dorsal raphe nucleus, in which four parts can be seen in this section, the dorsal (DRD), the ventral (DRV), the lateral (DRL) and the posterior (PDR) parts. (E) The substantia nigra. Two parts can be seen in this selected area, the compact part (SNC) and the reticular part (SNR). Note that a large number of neurons were immunolabeled in the SNC, whereas only a small number of neurons were immunolabeled in the SNR and their labeling intensity was lower. (F) The red nucleus, which can be divided into two parts, the parvicellular part (RPC) and the mangocellular part (RMC). Dorsal is top and lateral is left for A; dorsal is top and lateral is right for B, C, E and F; dorsal is to the up for D. Scale bar in F: 200 μm. (Modified from Sukiasyan et al., 2009).

Some modulatory nuclei including those containing monoaminergic cells such as substantial nigra and different raphe nuclei in the midline region (e.g., the dorsal raphe, median raphe and raphe obscures) were also intensely labeled. One nucleus that showed a strikingly high intensity of immunolabeling was the locus coeruleus, which may indicate that Cav1.3

mediates the efflux of noradrenaline from neurons in this nucleus [Sinnegger-Brauns et al., 2004; Striessnig et al., 2006]. Examples of the nuclei with a relative high Cav1.3 immunolabeling intensity in different regions of the brain stem are illustrated in Fig. 10.

6) Cerebellum

In the rat cerebellum, Cav1.3 has been revealed to be strongly expressed in the granular layer in the ISH and IHC studies [Hell et al., 1993; Tanaka et al., 1995; Chung et al., 2000]; however there existed discrepancies regarding its expression in Purkinje cells. Hell et al. [1993] and Grunnet and Kaufmann [2004] reported the clear presence of Cav1.3-IR in Purkinje cells whereas Chung et al. [2000] only detected a very weak labeling in these cells. In the molecular layer, cell bodies of basket and stellate cells are immunoreactive to Cav1.3 [Hell et al., 1993; Chung et al., 2000]. Cells in the cerebellar nuclei, which include the medial, interposed, and lateral cerebellar nucleus, were also Cav1.3 positive [Chung et al., 2000]. Based on all available published data, it seems that in the cerebellum the dendrites of different cell types do not contain a detectable amount of Cav1.3 channel proteins.

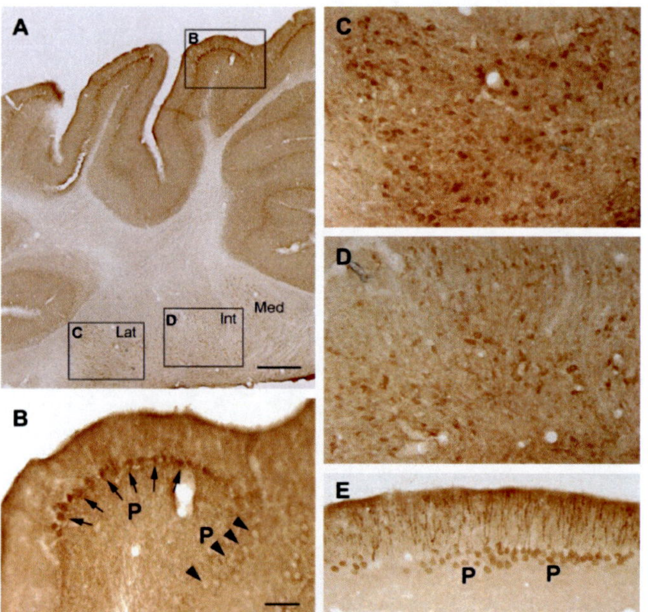

Figure 11. Transverse sections showing the Cav1.3 immunoreactivity in the cerebellum. (A) Low-power photomicrographs showing the Cav1.3 immunoreactivity in the cerebellar lobes and nuclei from a transverse section. (B) Enlargement of an area demarcated in (A) showing the detailed Cav1.3 immunolabeling pattern the cerebellar lobes. Note that the Purkinje cells (P) lining at the surface of the lobes were densely immunolabeled (arrows), whereas those in the deep part were weakly to negatively immunolabeled (arrowheads). (C, D) Enlargement of an area demarcated in (A) showing the detailed Cav1.3 immunolabeling pattern in the lateral (Lat) and interposed (Int) cerebellar nuclei. (E) Cav1.3 immunolabeling pattern on Purkinje cells (P) using another Cav1.3 antibody (see text). Note that by using this antibody the dendrites of the Purkinje cells (P) were clearly immunolabeled. Med: medial cerebellar nucleus. Scale bar in A: 500 µm; in B (for B – E), 100 µm.

We have immunolabeled a few series of cerebellar sections from two rats using a Cav1.3 antibody from Sigma-Aldrich (see Table 1) and found that the Purkinje and granular cells in the cerebellar cortex and the cells in the cerebellar nuclei were immunolabeled (Fig. 11). An interesting finding was that the intensity of immunolabeling of the Purkinje cells was decreased gradually from the surface of the cortex to the base of the fissures, i.e., those on the surface of the cerebellar cortex displayed a higher intensity of immunolabeling than those located in banks of the fissures, where some cells, especially those in the base of the fissures, even displayed negative immunolabeling (Fig. 11B). This phenomenon could also be noticed in the study by Grunnet and Kaufmann [2004]. In our study, the dendrites of Purkinje cells were not discernibly immunolabeled with the antibody from Sigma-Aldrich (Fig. 11B). However when we tried another Cav1.3 antibody obtained recently from Millipore-Chemicon, we found that it resulted in a clear immunoreactivity of dendrites of the Purkinje cells (Fig. 11E). The possible reasons underlying the diversity in the distribution of Cav1.3-IR will be discussed below.

3.   The spatial distribution of Cav1.3 in spinal motoneuron dendrites

The spatial distribution of Cav1.3 channels on neuronal soma and dendrites has been a controversial issue, largely due to the inconsistency of IHC data regarding the spatial distribution of Cav1.3-IR in spinal neurons in different animal species. As described above, thus far using IHC technique it has been shown that in cats and rats Cav1.3-IR are distributed mainly on spinal motoneuronal somata and proximal dendrites, in mice mainly on the dendrites, especially the second and third-order dendrites and in turtles presumably on the entire soma-dendritic surface. Given that many physiological studies have demonstrated that PICs were initiated mainly from motoneuron dendrites [Gutman, 1991; Hounsgaard and Kiehn, 1993; Delgado-Lezama et al., 1999; Svirskis et al 2001; Heckman et al., 2003, 2005] it is reasonable to expect that the channel proteins are also localized on the dendrites. These controversial IHC results encouraged further studies using other techniques, such as computer simulations, to investigate the functional spatial distribution of this channel [Elbasiouny et al., 2005; Bui et al., 2006; Grande et al., 2007]. The morphological and physiological parameters on which these simulation studies were based all originate from cat motoneurons. Unfortunately the results from these simulation studies were also inconsistent. Elbasiouny et al. [2005] concluded that Cav1.3 channels were primarily localized to a wide intermediate band overlapping with the dendritic territory for the muscle spindle Ia synapses at distances of 300-850 μm from the soma. Bui et al. [2006] suggested that Cav1.3 channels may be segregated to discrete hot spots 25-200 μm long and centered in the dendritic tree 100-400 μm from the soma. At the same time a study from the same research group also suggested that the location of the hot spots varied with the motoneuron size with the hot spots distributed further away from the soma in the larger neurons [Grande et al., 2007]. It is difficult to relate the results from these simulation studies to the physical anatomical distribution of the channel because what the simulation studies showed were "hot spots" of active channels and not all the channels in different functional states. The best way to reveal the physical distribution of the channels along the soma dendritic axis is to perform colocalization studies of channel immunoreactivity on intracellularly labeled neurons. However the outcome from this kind of study is also dependent on the quality of the antibody, the sensitivity of IHC technique and the method used to analyze the images. Ballou et al. [2006] have analyzed Cav1.3 distribution on fluorescence dye-labeled cat motoneurons with a program written for Matlab. They found that Cav1.3 was strongly expressed on the dendrites on a compartment near the cell soma, weakly and absently expressed 100-300 μm away from the soma and at least with one hot spot 500-1000 μm away from the soma. We have also performed a colocalization study on Neurobiotin-labeled cat motoneurons using a custom-made Cav1.3 antibody (provided by Dr. Bezprozvanny, University of Texas) [Zhang et al., 2008b] and our findings were different from Ballou et al. [2006]. Qualitatively the Cav1.3-IR could be seen along the entire length of the soma-dendritic axis with the highest signal in the soma. Along the dendritic tree the immunolabeling signal was higher in the proximal part and then gradually declined towards the distal parts (Fig. 12). Using a quantitative analysis using AutoQuant X (Media Cybernetics Inc., Bethesda, MD, USA) we have confirmed our qualitative observations (data not shown).

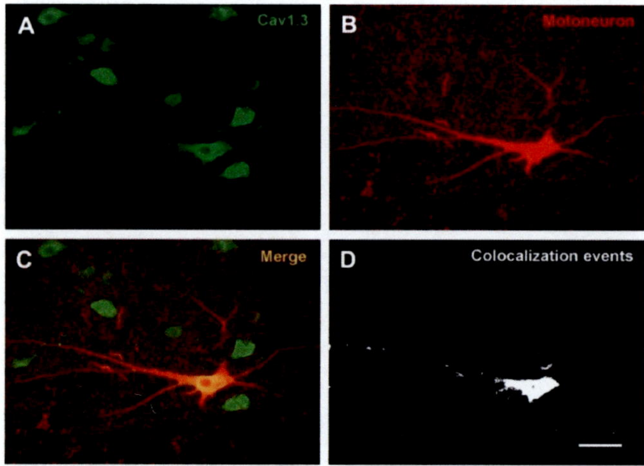

Figure 12. Cav1.3 immunolabeling on labeled cat motoneurons intracellularly labeled with Neurobiotin. (A) Cav1.3 immunoreactivity (green) in the ventral horn of a cat spinal section. (B) A motoneuron intracellularly labeled with Neurobiotin (red). (C) Merged image of A and B. (D) Colocalization events over the setting threshold processed with AutoQuant X program. It is clear that the neuronal somata and the proximal dendrites contained more colocalization events. Bar in D: 100 μm. (Adopted from Zhang et al., 2008b).

As stated above, one possible reason for the differences between the simulation studies and the IHC data is that the different studies may have detected different isoforms (splices) of Cav1.3 in the neurons. Using electron microscopy, we have demonstrated that Cav1.3-IR was located both in the neuronal somata and dendrites of different sizes [Zhang et al., 2008a]. We did not perform an electron microscopy study for the channel membrane distribution relative to the size of the dendrites on the cat motoneurons. However, evidence from rat globus pallidus and hippocampal neurons have showed that density of membrane-bound calcium channels (including Cav1.3) was greater on distal dendrites [Hanson and Smith, 2002; Leitch et al., 2009]. Further studies are needed to determine whether this spatial distribution pattern of Cav1.3 channels is the same on the dendrites of spinal motoneuron.

4. Cav1.3 expression changes in pathophysiological situations

In humans, it has been reported that structural aberrations within the pore-forming $\alpha1$ subunits of LTCCs can result in hypokalemic periodic paralysis and malignant hyperthermia sensitivity (Cav1.1), incomplete congenital stationary night blindness (Cav1.4), and Timophy syndrome (Cav1.2) [Striessnig et al., 2004, 2010]. Although the studies from Cav1.3 knockout mice suggested that the channel loss might affect sinoatrial node function and hearing [Platzer et al., 2000; Clark et al., 2003; Mangoni et al., 2003], Cav1.3 $\alpha1$ mutations have not been reported in humans until very recently when Baig et al. [2011] described a human Cav1.3 channelopathy (termed SANDD syndrome, an abbreviation of sinoatrial node dysfunction and deafness) with a cardiac and auditory phenotype that closely resembles that of Cav1.3 knockout mice [Platzer et al., 2000]. Because Cav1.3 channels also play an important role in the nervous system, we would like to know whether in some pathophysiological situations, for instance spinal injury, this channel is subject to any changes in levels of expression. It is known that after spinal cord injury PIC properties in motoneurons were lost in the acute phase due to the loss of the brain stem monoaminergic innervation, however the PICs reappeared 2-3 weeks after injury [Eken et al 1989; Bennett et al., 1999, 2004; Li et al., 2004 a,b]. Because Cav1.3 channels are believed to mediate PICs in spinal cord motoneurons, one may wonder whether these channels in motoneurons undergo a change in the levels of expression after spinal cord injury. Results from related IHC and gene expression studies were a bit surprising because neither at mRNA nor at protein level did the channel $\alpha1$ subunit show a significant upregulation in motoneurons of chronically spinalized rats [Anelli et al., 2007; Zhang et al., 2007; Wienecke et al., 2010]. Although the temporal changes in levels of Cav1.3-IR in spinal motoneurons of spinalized rats showed a slight down-regulation at 2 days and thereafter a slight up-regulation, the changes did not reach a significant level (Fig. 13) [Zhang et al., 2007].

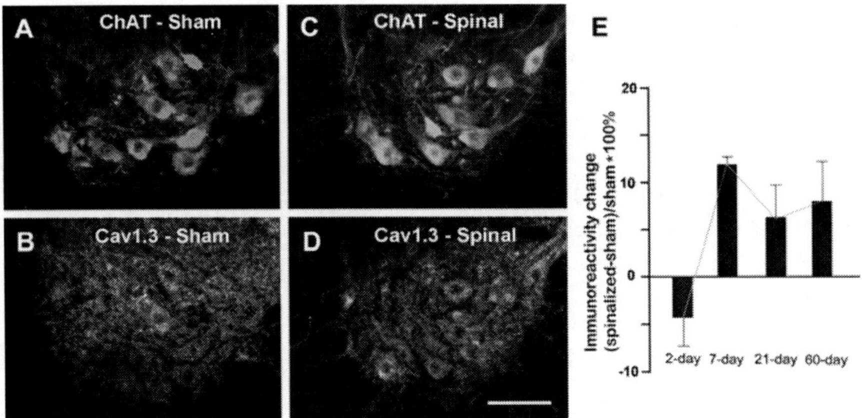

Figure 13. Cav1.3 immunoreactivity changes after spinalization in rats. (A-D) Photomicrographs from 60 day sham (A and B) or spinalized (C and D) rat sacral spinal cords immunolabeled with ChAT (A and C) or Cav1.3 (A and B) antibody. (E) Histogram showing the quantitative analysis of Cav1.3 immnuoreactivity changes in 4 different animal groups at different post-operation intervals (2, 7, 21 and 60 days). Although the Cav1.3 immunoreactivity was decreased by approximately 4% at 2 day following spinalization and increased by approximately 10% in the three late time intervals, the differences did not reach a significant level in any group. Bar in D: 100 µm.

However some auxiliary subunits of calcium channels, such as ß1, ß2, and ß4, showed a significant expression change (down- or up-regulation) either at protein or mRNA level [Zhang et al., 2007; Wienecke et al., 2010], but it is hard to judge which α1 subunit(s) they regulate. Dobremez et al. [2005] have studied Cav1.3 expression changes in rat spinal dorsal horn neurons after nerve injury (axotomy or ligation of sciatic nerve) both at mRNA and protein levels. They found that Cav1.3 was significantly down-regulated at 7 days but not at 3 days after nerve injury. All the data above indicate that an upregulation of the Cav1.3 α1 subunit *per se* is probably not an important factor for the enhanced excitability of spinal neurons after spinal injury.

5. Antibody specificity issue

By comparing the Cav1.3 immunolabeling pattern in the spinal cords of the different animal species we can see that conclusions regarding the distribution and cellular localization of Cav1.3 channels have proved somewhat controversial. There may be several possibilities to explain the discrepancies. The *first* possibility is that physical anatomical distribution of Cav1.3 channels is different between the different animal species. The *second* is that the antibodies from different sources label different channel isoforms (splices). The *third* is that the antibody used is not specific for the protein in the animal species investigated. The *first* possibility may hold true if the same antibody immunolabels different patterns on the same sort of tissues from different animal species; but the prerequisite is that the antibody must be validated for all the animal species. The *second* possibility is most likely true because Cav1.3 proteins may have many different splices owing to the alternative splicing property of LTCC α1 subunit [Ihara et al., 1995; Safa et al., 2001; Lipscombe et al., 2002]. Alternative splicing is one of the important properties of many LTCC α1 subunits, which increases the functional variations for specific cellular activities in response to changing physiological signals [Perez-Reyes et al., 1990; Hui et al., 1991; Snutch et al., 1991; Takimoto et al., 1997, Fan et al., 2005; Liao et al., 2005; Singh et al., 2008]. It is believed that a human calcium channel α1 gene may contain at least 10 sites of alternative splicing. If each site is independently regulated, alternative splicing could give rise to over 1000 distinct mRNAs from each α1 gene [Lipscombe et al., 2002]. This indicates that polyclonal antibodies from different batches may preferentially recognize different channel splices (isoforms). The immunoglobulins for a polyclonal antibody are obtained from different B lymphocytes in vivo. Polyclonal antibody preparations are usually a mixture of antibody specificities which all recognize the same antigen. The specificity difference means the antibodies bind with differing strengths to different epitopes on the antigen. This means that although the host animals are immunized with a specific immunogen the antibodies produced from different cell lines may differ and thus the quality is hard to keep identical from batch to batch. Our experience with Cav1.3 antibodies from Millipore-Chemicon favors this argument. As shown in Fig. 4 we have immunolabeled the mouse spinal cord using Cav1.3 antibodies from two different batches purchased from different times which were produced from the same immunogen (see Table 1). It is clearly seen that one antibody labeled predominantly the neuronal somata and proximal dendrites and the other labeled densely the dendrites of different sizes. This indicates that the antibody from one batch may be produced from one or several cell lines and recognizes Cav1.3 channel splices that are mainly located on the soma and proximal dendrites and that from the other batch may be produced from other cell lines and mainly recognizes

Cav1.3 channel splices that are located on the dendrites. In such a case we cannot just simply say that one antibody is specific and the other one is not. The *third* possibility can definitely not be excluded because the antibody validation is indeed not a trivial issue. There are many papers to discuss how to verify the antibody specificity [e.g., Saper, 2005, 2009; Rhodes and Trimmer, 2006] and some journals like *the Journal of Comparative Neurology* requests a strict antibody specificity control for the studies that involves antibodies. Several antibody control methods have been proposed, which include immunostaining the specific gene knockout animal tissue with the antibody, adsorbing the antibody with the antigen and western blot analysis, etc. [see Saper, 2005]. However one important and easily ignored issue is that before doing IHC on a specific species one must make a search of aimed protein sequence from a protein bank and make sure that the epitope is in the sequence that the antibody recognizes. For instance, as far as we know Cav1.3 channel sequence of the turtle (*Chrysemys scripta elegans*) is unavailable right now, instead it has been tacitly assumed that its sequence is the same as that in the rat and thus the antibodies produced with rat Cav1.3 antigen were used. In such a case using knockout mouse tissue or adsorption control are not appropriate. We have tried Cav1.3 antibodies produced using antigen either from rat or human and found totally different immunolabeling patterns on the turtle spinal cord (Fig. 5), although preadsorption of the antibody with the antigen peptide provided by the manufacturer always produced a negative immunostaining result. Unfortunately the preadsorption control is meaningless for monoclonal antibodies and for polyclonal antibodies that have already been affinity purified [Saper, 2009]. Thus, to solve such a dilemma we have to wait until a specific channel gene of the animal of interest, e.g., turtle, is cloned.

COMMENTS ON THE FUNCTION OF CAV1.3 CHANNELS IN CNS NEURONS

The electrophysiological data available regarding the function of Cav1.3 channels in CNS neurons are mainly related to their voltage-sensitive response to classical ionotropic excitatory postsynaptic potentials (EPSPs) generated in their neighborhood. The properties of the Cav1.3 channels will thus *amplify* and *prolong* the response by the EPSPs themselves. Another aspect is the interaction with neighboring voltage (or Ca^{2+}) sensitive K^+ channels which can make the Cav1.3 channels a pivotal part of the pacemaker properties of central neurons. In contrast there are only few publications which are related to the possible action at presynaptic terminals in relation to synaptic transmission, or a role in activating signaling pathways controlling gene expression.

For a general discussion on the function of the Cav1.3 channels in CNS neurons it may useful to give an account of the initial observations that lead up the conclusion that these channels play an important role in information processing in spinal motoneurons.

1. Function of the Cav1.3 channel in spinal motoneurons

The mammalian motoneuron was historically thought of as a simple summation device; unitary postsynaptic excitatory and inhibitory potentials from thousands of synapses over the soma-dendritic membrane were thought to just spread passively across the membrane [Rall,

1992; Eccles, 1957; Granit, 1972]. With the lowest threshold for the action potential localized to the initial segment / axon hillock (due to the high concentration of the sodium ion (Na^+) channels there) this would be the "summating point". It was therefore something of a paradigm shift in the research on motoneurons with Schwindt and Crill's discovery of a low threshold voltage sensitive non-inactivating (persistent) inward current in cat spinal motoneurons in the late 1970ies [Schwindt and Crill, 1977, 1980]. Fig. 14 A and B illustrates the results from their first "two electrode" voltage clamp experiments in anaesthetized cats. The motoneuron was thus penetrated by two independent microelectrodes, one for controlling the voltage, the other for injecting and measuring the current needed to produce the preset voltage ramp. With this experimental design (Fig. 14A and B) they described a region of "negative slope conductance" in the current–voltage relation of motoneurons [Schwindt and Crill, 1977]. These results indicate that at a membrane potential at around -50 mV there are voltage sensitive ion-channels that are opened, which conduct positively charged ions into the cell; thus less external current injection was needed to force the predetermined depolarization to occur. The authors speculated at that time that this PIC could have a role in the amplification of synaptic inputs. They also described that when this current is large enough it can support long-lasting discharges (in current clamp mode), far outlasting the excitatory input.

Subsequently, Hultborn and colleagues used intracellular recording techniques to demonstrate that the prolonged firing in response to transient reflex activation indeed was produced intrinsically in the motoneuron by an all-or-none plateau depolarization [Hounsgaard et al., 1984]. They found that such a response could be produced not only by stimulating sensory afferents but also by stimulating motoneurons themselves with intracellular depolarizing current injection into the soma (Fig. 14C). The plateau was similarly terminated by a hyperpolarizing current. When injecting a triangular (depolarizing–repolarizing) current pulse, they further noted that the frequency–current relation was hysteretic (Fig. 14D and E), i.e. the firing rate during the repolarizing ramp was higher than at the equivalent current during the depolarizing ramp. Fig. 14F and G illustrates more directly the changes in membrane potential when the spike potentials were later inactivated. An additional important observation was that these intrinsic properties of the motoneurons depended on the presence of the neuromodulator serotonin (5-HT) [Hounsgaard et al., 1988].

It was not easy to analyze the ionic mechanisms of the plateau potentials (by the PIC) further in the cat *in vivo* preparation. As the 5-HT–dependent plateau potentials in motoneurons could be reproduced under *in vitro* conditions in the isolated turtle spinal cord [Hounsgaard et al., 1985] the road was open for the Hounsgaard team to analyze the ionic components of the PIC as well as the voltage sensitive channels involved in the response. The early results from the *in vitro* turtle preparation demonstrated that the "plateau current" was conducted by Ca^{2+} (TTX did not affect the size of the plateaux in the turtle – thus a significant Na^+ contribution was unlikely in that species). As for the type of ion channel they [Hounsgaard and Mintz, 1988; Hounsgaard and Kiehn, 1989] demonstrated that the current could be blocked by nifedipine, thus pointing to the LTCCs. Since the activation of these "plateau currents" occurred at rather hyperpolarized levels (even below the spike-threshold) it was concluded that the effects were originated from the Cav1.3 subtype which had been shown to be activated at more hyperpolarized levels than the Cav1.2 subtype [Perrier et al., 2002].

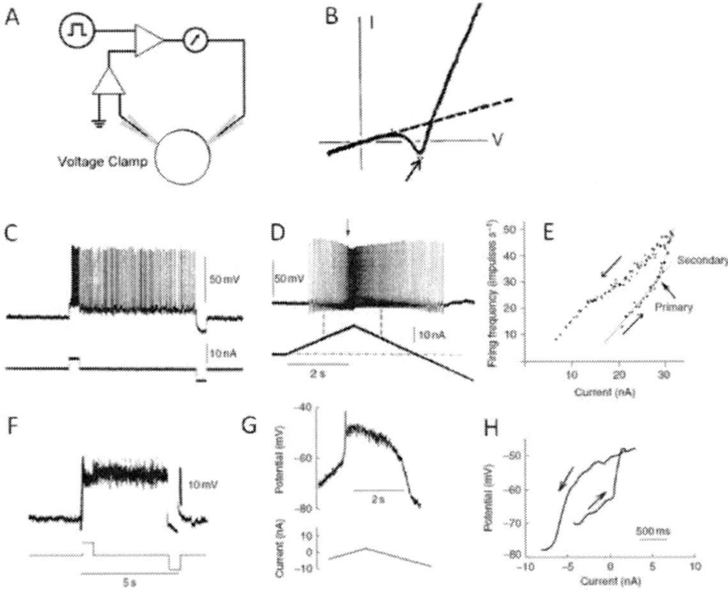

Figure 14. Persistent inward currents and plateau potentials in spinal motoneurons. A-B: Voltage clamp recordings from a spinal motoneuron with a "negative slope conductance" produced by a voltage dependent non-inactivating inward current. The thin line represents the current expected to be needed to depolarize the motoneuron with a passive membrane. However, at ~-50 mV much less current is needed because of the activation of a persistent inward current; at more depolarized levels more current has to be injected due to activation of K^+ conductance. C-H: Signs of plateau potentials recorded in spinal motoneurons; sustained shifts in excitability triggered by depolarizing and hyperpolarizing currents injected intracellularly into cat spinal motoneurons in a decerebrate, unanesthetized preparation. C. Sustained repetitive firing in a motoneuron, which was initiated by a short depolarizing current pulse and terminated by a short hyperpolarizing pulse; D. Firing pattern and membrane potentials during injection of triangular current pulses (spike activity in a motoneuron (upper trace) and triangular profile of its injected current (second trace); E. Graph of D; E. Sustained depolarization in a motoneuron following an injection of QX 314 to block Na^+ channels; F. membrane potential during injection of triangular current pulses (showing the membrane potential without spikes); G. graph of F. In C-D and F-G, the upper traces are intracellular recordings and the lower traces indicate the amount and timing of the injected current. In D the interrupted vertical lines point to the difference in firing frequency at similar current intensities during the ascending and descending phase; the interrupted horizontal line shows the zero level for injected current. In the graph in C, the thin lines approximate the f /I slopes for Kernell's primary and secondary ranges of firing. Note the counterclockwise hysteresis. (B: modified from Schwindt and Crill, 1977, with copyright permission from Elsevier; C-H: modified from Hultborn and Fedrichuk, 2009).

The continuation of the research on the PIC and plateau properties in spinal motoneurons now includes many laboratories around the world and several recent reviews cover different aspects of the plateau potentials in spinal motoneurons [Powers and Binder, 2001; Heckman et al., 2003, 2005; Hultborn et al., 2004; Brownstone, 2006; Heckman et al., 2009; ElBasiouny et al., 2010].

Although the opening of Cav1.3 channel is voltage-dependent, it is also controlled by a number of neuromodulators. Initially the facilitation of plateau potentials in motoneurons by

5-HT was tentatively explained by a 5-HT-mediated reduction of outward K^+ currents, thus "uncovering" the plateau current [Hounsgaard and Kiehn, 1989]. However, subsequent studies by Hounsgaard and colleagues in the turtle have shown that several other transmitters (via G-protein-coupled receptors) may be involved and that the effect is also due to a facilitation of the LTCCs [Delgado-Lezama et al., 1997; Svirskis and Hounsgaard, 1998; Powers and Binder, 2001; Perrier et al 2002; Alaburda et al., 2002; Hornby et al., 2002]. The Hounsgaard group has also demonstrated that anesthetics such as barbiturates suppress the LTCCs [Guertin and Hounsgaard, 1999].

For a further interpretation of the functional significance of the plateau properties it would be important to know the soma-dentritic location of the plateau current. As described above, the immunohistochemical evidence is not easy to interpret as the results seem to differ for different antibodies – perhaps explained by their relative specificity to the various splice variants that may vary with the localization of the Cav1.3 channels. The electrophysiological evidence favors that the plateau current is predominantly of dendritic origin. This was first predicted on the basis of a theoretical analysis of the data of Schwindt and Crill [Gutman, 1991]. This prediction was later supported by experimental analysis of the dynamics of the hysteresis in the plateau current caused by the dendritic cable properties for the turtle motoneurons [Svirskis et al., 2001], and later for mouse motoneurons [Carlin et al., 2000a]. The dominant dendritic origin was also more directly demonstrated in experiments on turtle motoneurons in which the soma and dendrites were differentially polarized by application of extracellular electrical fields; plateau potentials could be triggered with a depolarization of the dendrites even when the soma region (with the recording microelectrode) was hyperpolarized [Hounsgaard and Kiehn, 1993].

Are plateau potentials "all-or none" in character, or are they graded under physiological conditions? Mostly the experimental paradigms have employed current injection through a recording electrode that was presumably positioned in the cell soma. This current will thus affect the whole motoneuron (albeit with decreasing amplitude in the distal dendrites due to the electrotonic cable properties) and may have had a tendency to overemphasize the 'all-or-none' character of the current producing the plateau potential. It seems functionally more realistic to imagine that a local synaptic excitation (predominately in the dendrites) would activate a PIC only in its immediate environment – the local PICs would then boost/amplify this synaptic excitation, although it would be insufficient to generate a full-blown plateau potential involving the whole neuron. Indeed, electrophysiological experiments in the turtle [Delgado-Lezama et al., 1999], and the adult cat [Hultborn et al., 2003] all support the notion that the Cav1.3 PICs can cause a large and local amplification of synaptic excitation that is graded for the motoneurons as a whole.

Is the entire "plateau current" in the motoneurons mediated by Cav1.3 channels? Certainly the higher threshold portion by Cav1.2 channel may be recruited as the depolarization caused by the current through the Cav1.3 channel reaches the Cav1.2 threshold, but another important question is whether also non-inactivating Na^+ channels contribute. This question can hardly be answered in *in vivo* experiments, but *in vitro* experiments on the sacrocaudal rat spinal cord has revealed a significant TTX-sensitive part of the PIC (thus mediated through a Na^+ channel, see e.g., [Harvey et al., 2006]). In the chronic spinal rats (transection at the S2 segment causing spasticity in the tail muscles), it was revealed that the Ca^{2+}- and Na^+-PICs contributed almost equally in spinal motoneurons during the initial phase of PIC, whereas Ca^{2+} contributed to approximately 2/3 of the total

PICs at the sustained peak [Li and Bennett, 2003]. Thus the Ca^{2+}-PICs seem to inactivate with a much slower time course than the Na^+-PICs. Recently it was reported that also the turtle motoneurons indeed do have a significant Na^+ component [Gabrielaitis et al., 2011].

2. Functional significance of the Cav1.3 channel in the spinal cord - interneurons and ascending neurons

Plateau potentials or PICs have been reported in neurons in all parts of the rat spinal cord gray matter, including the dorsal horn and the ventral horn in the rat [Morisset and Nagy, 1998, 1999; Li et al., 2004a; Harvey et al., 2006; Theiss et al., 2007] as well as in the turtle [Hounsgaard and Kjaerulff, 1992; Russo and Hounsgaard, 1996]. The widespread presence of the Cav1.3 immunolabeled neurons in different laminae of the gray matter [in the rat: Sukiasyan et al., 2009] may be partly related to these findings. The "plateau properties" have been recorded either as plateau potentials, prolonged firing with lower current intensities at the descending phase of triangular current injections in current clamp mode, or as an inward current in voltage clamp mode. As the experimental approaches have been very different in the various investigations it is difficult to achieve a direct comparison between the degree of "plateau properties" on one side and the intensity of Cav1.3 labeling on the other. The comparison is further hampered by the contribution of Na^+-currents to the plateau properties (as discussed above). Nevertheless there seems to be a certain correlation between the two observations as discussed below. As expected from the intense labeling of neurons (mainly motoneurons) in the lateral part of the ventral horn, a large proportion of motoneurons have been shown to have "plateau properties" (*in vitro*: both in acute preparations following administration of monoamines, or in the chronic spinal state without additional monoamines, *in vitro*: [Li et al., 2004a]; *in vivo*: [Button et al., 2006]). In a recent study using spinal cord slices from 11-19 days old rats Theiss et al. [2007] it was demonstrated that ventral horn interneurons could be classified in four groups according to their firing pattern: repetitive-firing, repetitive/burst, initial burst or single spiking. Among these, a large proportion (62%, 26/42) of the sampled repetitive-firing neurons had significant PICs, whereas the percentage was lower for the other three types of interneurons. In total approximately about 46% (32/70) of all the sampled ventral horn interneurons showed significant plateau properties. In the deep dorsal horn only about one third (28-35%) of its interneurons could generate plateau potentials [spinal cord slices, *in vitro*: Morisset and Nagy, 1996, 1998].

It seems that the contribution of the persistent Ca^{2+} and Na^+ channels to the total PICs varies with the neuronal localizations in the rat spinal gray matter. Thus in the ventral horn interneurons the largest portion of the total PIC was contributed by the persistent Na^+ current, whereas the persistent Ca^{2+} current contributed only about 12% of PICs on the ascending phase and 45% on the descending phase with triangular voltage commands [Theiss et al., 2007]. In the deep dorsal horn neurons only the Ca^{2+}-PICs were studied, so information on the possible contribution of Na^+-PICs is not available [Morisset and Nagy, 1996, 1998].

Neurons in some specific sensory neuronal groups with ascending projections, like the lateral cervical nucleus, the lateral spinal nucleus, the central cervical nucleus, and Clarke's column [Matsushita and Hosoya, 1982; Giesler et al., 1988; Burstein et al., 1990], were moderately labeled [cat: Zhang et al., 2008a; rat: Sukiasyan et al., 2009]. Nevertheless, in the cat there has been no signs of plateau properties in Clark's column cells (dorsal spinocerebellar tract neurons) although they have rather strong Cav1.3 immunolabelling, and neighboring motoneurons in the same experiments did show plateau potentials [Stecina, Zhang and Hultborn, unpublished data]. The plateau properties of the other ascending projecting neurons in these nuclei have not been studied so far, neither in the cat, nor in the rat. Both in the cat [Zhang et al., 2008a] and the rat [Sukiasyan et al., 2009] the motoneurons in the Onuf's nucleus were strongly labeled. The presence of plateau potentials in this nucleus was reported in cat [Paroschy and Shefchyk, 2000], but electrophysiological studies in the rat are still lacking.

3. Functional significance of Cav1.3 channels for motor and sensory neurons in the brain stem

The ability to generate PICs or plateau potentials has been reported for several brain stem motor nuclei, including the trigeminal, facial and hypoglossal motor nuclei [Hsiao et al., 1998; Powers and Binder, 2003; Cramer et al., 2007; Moritz et al., 2007]. The mechanisms for the generation of PICs in the neurons in the brain stem seemed to be more complicated than in the spinal cord. As for the spinal cord, persistent Na^+-currents also contribute to the total PICs observed in the brain stem nuclei [Sandler et al., 1998; Powers and Binder, 2001; Cramer et al., 2007]. For the hypoglossal motoneurons it was also demonstrated (pharmacologically) that not only the LTCCs but also other type (e.g., P- and N-type) calcium channels contribute to the Ca^{2+}-PICs [Powers and Binder, 2003]. Nonetheless, a recent study from the same group using the "nucleated patch" technique showed that the major portion of the Ca^{2+}-PICs and the prolonged tail currents were carried primarily by somatic Cav1 channels [Moritz et al., 2007]. The PICs in the facial motoneurons (innervating the mystacial pad muscles) were mainly mediated by a persistent Na^+ current (~80%), but there is also small portion that is probably mediated by LTCCs [Cramer et al., 2007]. In the presence of 5-HT a bistable membrane behavior has been found in the trigeminal motoneurons from mature guinea pigs [Hsiao et al., 1998]. This behavior was mediated by both persistent Na^+ and LTCCs and the latter seemed more important because bistable behavior continues in the presence of TTX. In conclusion, bistable behavior is a property not only for spinal motoneurons but also for brain stem motoneurons. The strong Cav1.3 labeling intensity in the neurons in the brain stem motor nuclei may imply that this channel provides an important contribution to the total Ca^{2+}-PICs in these nuclei.

There are relatively few studies on the plateau potentials in the brain stem sensory neurons. However, at least two studies on the trigeminal principle sensory nuclei did show that bistability exists in the neurons in this nucleus in either physiological or pathological conditions. One study was performed on young gerbils using the *in vitro* whole-cell patch recording technique [Sandler et al., 1998]. The results showed that a small portion of investigated trigeminal principle sensory neurons have the ability to generate plateau potentials and that the plateau properties were mediated mainly by Na^+ channels (the plateau potential could be induced in the Ca^{2+} free medium and could be eliminated by application of TTX). The other study investigated the bistable properties of deafferented barrelette cells in the trigeminal principle sensory nuclei in young rats (8 days after inferior orbital nerve cut at birth) using the same recording technique [Lo and Erzurumlu, 2002]. The results showed that almost all of the investigated neurons (19/20) exhibited bistable behavior and that these plateau potentials could be completely blocked by L-type Ca^{2+}-channel antagonists. Under normal conditions (no lesions), there were no signs of plateau properties for the neurons in this nucleus [Lo et al., 1999], but there has been no trials in which various transmitters (e.g., monoamines) were added to facilitate the plateau behavior.

4.　Functional significance of Cav1.3 channels in monoaminergic neurons in brain stem

One group of nuclei with strong Cav1.3 labeling is those containing monoamines as transmitter [Takada et al., 2001; Sukiasyan et al., 2009]. These nuclei include, the locus ceuroleus (noradrenergic), most of the raphe nuclei (serotonergic), and a number of dopaminergic nuclei [A1 – A17, Ungerstedt, 1971; Hökfelt et al., 1984], among those the compact part of the substantia nigra [Felten and Sladek, 1983]. One common electrophysiological characteristic of these neurons is their high excitability with a tendency to fire tonically spontaneously and rhythmically [e.g., Williams et al., 1984; Hainsworth et al., 1991; Hajós et al., 1995]. Although it seems that most research on the mechanism underlying this tonic spontaneous firing implicates Cav1.3 channels, it is still unclear whether their function is related to the increase in intracellular Ca^{2+} (and subsequent opening of Ca^{2+} - dependent K^+ channels [Wilson and Callaway, 2000; Vandael, 2010], or whether it is the voltage gated persistent current *per se* [Putzier et al., 2009]. In a recent review Vandael et al. [2010] discussed the evidence for a pacemaker function for Cav1.3 channels for 3 specific cell types: the dopaminergic neurons in compact part of the substantia nigra, the cardiac pacemaker cells and the adrenal chromaffin cells. In the case of the neurons in compact part of the substantia nigra the direct evidence originates from *in vitro* studies using slices from the rat brain stem [Kang and Kitai, 1993; Nedergaard et al., 1993; Mercuri et al., 1994; Wilson and Callaway, 2000; Puopolo et al., 2007; Putzier et al., 2009]. In the rat the locus ceuroleus nucleus is one of the areas with the most dense Cav1.3 immunolabeling [Sukiasyan et al., 2009], while the labeling is distinctly weaker in the mouse [Sukiasyan, Hultborn and Zhang, unpublished data]. Related to this possible species difference in Cav 1.3 immunolabeling, electrophysiological analysis demonstrated a strong Ca^{2+} -component of the pacemaker potential during the interspike interval in the rat [Williams et al., 1984], while the Na^+ -component was dominating in the mouse [de Oliveira et al., 2010].

The brain stem raphe nuclei are the main origin of 5-HT fibers that project to many other central nervous structures. Among these are the ascending serotonergic fibers which originate mainly from pontine and midbrain raphe nuclei (dorsal raphe and median raphe; [Adell et al., 2002]) and the descending serotonergic fibers originating mainly from the medullary raphe nuclei (raphe obscurus, raphe pallidus, and raphe magnus; [Li et al., 1993]). The intrinsic origin and control of the tonic or rhythmic firing activity of the brain stem serotonergic neurons is not yet clear. The contribution of various glutamate receptors in the regulation has been specifically investigated [Gartside et al., 2007], while the role of the high concentration of Cav1.3 channels needs to be studied further.

5.　Functional significance of Cav1.3 channel in the basal ganglia

The medium spiny neurons in the striatum are strongly Cav1.3 immunolabeled [Olson et al., 2005; material presented in Fig. 7 in this chapter]. The medium spiny neurons are characterized of having two preferred membrane potentials (both subthreshold for action potential generation) referred to as the "down-state" and "up-state" [Wilson and Groves, 1981]. The mechanisms that give rise to preferred membrane potentials could either involve network properties (continuous excitatory input during the up-state [Wilson and Kawagushi, 1996]) or be due to intrinsic membrane properties [Vergara et al., 2003]. In a cortico-striatal slice preparation, Vergera et al. [2003] demonstrated that NMDA-induced bistability could be partly blocked by L-type channel antagonists. Olsen et al. [2005] later demonstrated that the "upstate transition" in the medium spiny neurons (driven by glutamate receptors) was suppressed in Cav1.3 knock-out mice. These data show that NMDA and L-type Ca^{2+} conductance of spiny neurons are capable of rendering them bistable. It therefore seems likely that the Cav1.3 PICs are essential for the up-state although plateau potentials have not been induced by intracellular current pulses as described for spinal motoneurons.

6. Functional significance of Cav1.3 channel in the cerebral cortex

As described above, cortical neurons throughout layers II-VI were often Cav1.3 positive in rats. However, investigations on PICs in cortical slices often emphasize the presence of TTX-sensitive PICs [Schwindt and Crill, 1995; reviewed by Crill, 1996]. These persistent currents have been assumed both to amplify synaptic input and control spike pattern, timing and frequency (see Crill, 1996; Storm et al., 2009), but have also been specifically emphasized that they may contribute to the generation of the oscillatory activity [Llinás et al., 1991].

There are numerous publications regarding "persistent activity" recorded in cortical neurons in behaving animals, in relation to short term memory and working memory [e.g. Goldman-Rakic, 1995, 1996; Wang, 2001]. However, most of these studies hypothesize the presence of specific closed loop circuits for maintaining this activity, although a few publications suggest that a metabotropic induction of plateau properties by nifedipine sensitive Ca^{2+}-channels could be involved [Zhang and Ségéla, 2010]. In most cortical neurons studied there are "spontaneous" shifts between DOWN and UP states where the cortical neurons are much more sensitive to additional synaptic inputs during their UP-state than during the DOWN-state. Although it was once thought that the UP- and DOWN-states depended on very large networks of neurons (and therefore could only be studied *in vivo*), Sanchez-Vives and McCormick [2000] demonstrated that these shifts remained in cortical slices *in vitro*. Some work has been specifically aimed at integrating the intrinsic cellular properties [e.g. McCormick, 2005], but the mechanism(s) remain elusive, and there is no direct evidence demonstrating that PICs are necessary for these shifts in membrane potential.

It is also possible that the Cav1.3 channels are more important for the control of Ca^{2+} entry in relation to long term plastic changes in relation to "memory and learning" in the case of cortical neurons. In line with this suggestion it has recently been demonstrated that Cav1.3 channels mediate the consolidation of contextually conditioned fear in wild type mice, since this consolidation was impaired in Cav1.3 knockout mice [McKinney et al., 2009]. That work builds on earlier evidence that Cav1.3 channels are essential for a form of "long term potentiation" in amygdala [Weisskopf et al., 1999].

7. Reverberating activity in closed neuronal circuits – a role for PIC as well?

Integrative functions of a neuronal network depend on both the connectivity between neurons as well as on the intrinsic neuronal properties of the participating neurons. Historically the concept of a "reverberating activity in closed neuronal circuits" was first proposed in 1922 by Forbes where he suggested that spinal reflex afterdischarges were produced by recirculation of impulses along postulated closed loops (circuits of neurons). The idea was subsequently extended by Lorente de No [1938] and Burns [1958]. There has been increased attention to the fact that voltage-dependent non-inactivating PICs could permit short-lasting inputs to produce long-lasting output acting as a "neuronal integrator", e.g. in the vestibulo-ocular reflex. Some recent work on neuronal integrators suggests that the stability of the "neuronal loop" performance would be enhanced by additional "bistable" intrinsic neuronal properties [Koulakov et al., 2002; Aksay et al., 2007].

Decades ago Tsukahara and collaborators [Tsukahara et al., 1971, 1983] described reverberating activity in a simple two-neuron loop, with the two nuclei located conveniently far apart to facilitate an experimental analysis. Fig. 15 illustrates the connectivity (A) and the electrophysiological evidence of mutual excitatory connections between the reticulotegmental nucleus of the pons and the interposed nucleus in the cerebellum. The neurons of the latter nucleus project to the neurons of the rubrospinal tract (and contribute the strongest excitatory input to these neurons) as well as to the ventrolateral thalamic nucleus. Single electrical stimuli to the cerebellar interposed nucleus (or their axons in the ventrolateral thalamic nucleus) or the reticulotegmental nucleus (or their connecting fibers) could result in a long lasting activity in this loop, mostly monitored by recording a long lasting depolarization from the neurons in the red nucleus. If the animals were anaesthetized, or had the inhibitory side-loop via the Purkinje cells intact, single stimuli only evoked a single monosynaptic EPSP in neurons in the red nucleus (Fig. 15B).

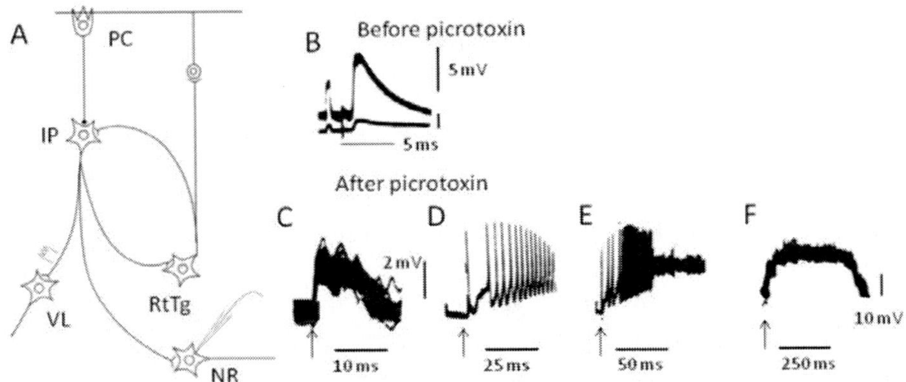

Figure 15. A. Diagram of the "closed loop" that has been demonstrated to support reverberating activity when the Purkinje cell (PC) inhibition is removed by picrotoxin. It has been shown that all neurons in the loop are strongly Cav1.3 immunolabeled, suggesting that intrinsic properties could contribute to the maintained activity in the loop. The two-neuron loop consists of reticulotegmental nucleus (RtTg) in the pons and the interposed nucleus (or nucleus interpositus, IP) in the cerebellum. The IP projects not only to RtTg but also to the red nucleus (RN) neurons and the ventrolateral thalamic nucleus (VL). The activity in the loop is monitored by recordings from the RN neurons (B-H), and the activity in the loop is initiated by stimulating the IP axons by single shocks close to the VL nucleus. B-E: Intracellular recording of EPSPs in a RN neuron following single stimulus shocks to the IP axons in VL. Before picrotoxin there is only a monosynaptic EPSP (an "axon reflex"), but after the picrotoxin the short latency (monosynaptic spike) is followed by a maintained firing that is supported by the closed loop. It seems likely, however, that PICs mediated by Cav1.3 channels could also support this activity, although this has not yet been investigated. (Modified from Tsukahara et al., 1971, with copyright permission from Elsevier).

In unanestehtized preparations (high decerebration) with cerebellar cortical ablation (or pharmacological application of picrotoxin) to prevent the Purkinje cell inhibition of the interposed cerebellar nucleus the intracellular recordings revealed ripples of EPSPs reflecting to the cerebello-pontine loop delay. These ripples were often seen superimposed on a maintained depolarization (Fig. 15C-F). The maintained activity could be prevented by cooling either the middle or inferior cerebellar peduncles, which mediates the mutual connections between the reticulotegmental and the interposed cerebellar nuclei [Tsukahara et al., 1983]. In their discussion they clearly pointed out that the maintained activity could be dependent both on "network properties" (closed loop with reverberate activity), and/or to intrinsic membrane properties – they themselves had provided evidence for mutual excitatory connections between the reticulotegmental nucleus and the interposed cerebellar nucleus. We have demonstrated that in the rat the neurons of the reticulotegmental nucleus are strongly Cav1.3-positive [Sukiasyan et al., 2009] and the same is true for the deep cerebellar nuclei [Chung et al., 2000; see also Fig. 11 in this chapter]. Finally the "output" neurons in the red nucleus, from which the maintained activity was usually monitored, are intensively labeled [Sukiasyan et al., 2009]. Thus it seems that anatomical connectivity and PICs (suggested by the Cav1.3 labeling) both contribute to a remarkable maintained activity following single electrical stimuli along the reticulotegmental – interposed nuclear loop. Hopefully the actual contribution by the intrinsic properties of the involved neurons will be directly tested in the future.

CONCLUSION

In this chapter we have presented the recent advances in research field of Cav1.3 channels with respects to their expression in the CNS and their possible physiological functions. Cav1.3 is an important LTCC, which is expressed pervasively across the CNS over different animal species, including the cat, rat, mouse and turtle. Although there exists difference in its regional expression among the different animal species, Cav1.3 was found to be expressed extensively in the spinal gray matter in different segments. However, when describing the different Cav1.3 immunolabeling patterns in the different animal species, one must keep in mind that the different immunolabeling patterns could also be a consequence of the different antibodies used in the different studies. Nevertheless, Cav1.3 in spinal motoneurons always had a higher level of expression than in other types of spinal neurons regardless of the animal species. Although a number of investigations have been performed more studies are still needed to reveal the spatial distribution of Cav1.3 channels on the dendrites of spinal motoneurons. The distribution of Cav1.3 channels in the brain has been mainly examined in rats and has shown that Cav1.3 channels were widely distributed across different regions in the brain, including the olfactory bulb, cerebral cortex, hippocampus, basal ganglia, diencephalon, cerebellum and brain stem. As observed in the spinal cord, the Cav1.3-IR in different brain regions (nuclei) was also not evenly distributed. In some regions (nuclei), e.g., the cranial motor nuclei in the brain stem and the nuclei contain the monoaminergic neurons Cav1.3 was expressed more intensely.

Except the general physiological functions stated in the Introduction, Cav1.3 may exert many other special functions in the CNS. In the spinal cord, it is strongly believed that Cav1.3

is related to the generation of the PICs in motoneurons. Such voltage-gated PICs have been assumed to boost and amplify the classical ionotropic EPSPs, especially when the PIC properties are facilitated by monoaminergic innervation. In several cases Cav1.3 channels have been implicated as driving the pacemaker potentials and thus causing tonic spontaneous firing. However it is still an open question as to whether this depends on the Ca^{2+} entry, or is dependent on the PICs *per se*. In a few cases the Ca^{2+} entry is part of the mechanisms underlying the consolidation of "memory and learning", thus indicating its possible role in the control of gene expression.

ACKNOWLEDGMENTS

We are grateful to Lillian Grøndahl for her help in all aspects for the studies in this work. We thank Claire Meehan for reading of the manuscript. We thank the following foundations for supporting our different research projects. They include the Ludvig & Sara Elsass´ Foundation, the Danish Medical Research Council, the Lundbeck Foundation and the Novo Nordisk Foundation and the Danish Multiple Sclerosis Foundation.

REFERENCES

Adell, A; Celada, P; Abellán, MT; Artigas, F. Origin and functional role of the extracellular serotonin in the midbrain raphe nuclei. *Brain Res Brain Res Rev,* 2002; 39:154-180.

Ahlijanian, MK; Westenbroek, RE; Catterall, WA. Subunit structure and localization of dihydropyridine-sensitive calcium channels in mammalian brain, spinal cord, and retina. *Neuron,* 1990; 4:819-832.

Aksay, E; Olasagasti, I; Mensh, BD; Baker, R; Goldman, MS; Tank, DW. Functional dissection of circuitry in a neural integrator. *Nat Neurosci,* 2007; 10:494-504.

Alaburda, A; Perrier, JF; Hounsgaard, J. Mechanisms causing plateau potentials in spinal motoneurones. *Adv Exp Med Biol,* 2002; 508:219-226.

Anelli, R; Sanelli, L; Bennett, DJ; Heckman, CJ. Expression of L-type calcium channel alpha(1)-1.2 and alpha(1)-1.3 subunits on rat sacral motoneurons following chronic spinal cord injury. *Neuroscience,* 2007; 145:751-763.

Arikkath, J; Campbell, KP. Auxiliary subunits: essential components of the voltage-gated calcium channel complex. *Curr Opin Neurobiol,* 2003; 13:298-307.

Ashcroft, FM; Proks, P; Smith, PA; Ammala, C; Bokvist, K; Rorsman, P. Stimulus-secretion coupling in pancreatic beta cells. *J Cell Biochem,* 1994; 55:54–65.

Baig, SM; Koschak, A; Lieb, A; Gebhart, M; Dafinger, C; Nürnberg, G; Ali, A; Ahmad, I; Sinnegger-Brauns, MJ; Brandt, N; Engel, J; Mangoni, ME; Farooq, M; Khan, HU; Nürnberg, P; Striessnig, J; Bolz, HJ. Loss of Ca(v)1.3 (CACNA1D) function in a human channelopathy with bradycardia and congenital deafness. *Nat Neurosci,* 2011; 14:77-84.

Ballou, EW; Smith, WB; Anelli, R; Heckman, CJ. Measuring dendritic distribution of membrane proteins. *J Neurosci Methods,* 2006; 156:257-266.

Baroudi, G; Qu, Y; Ramadan, O; Chahine, M; Boutjdir, M. Protein kinase C activation inhibits Cav1.3 calcium channel at NH2-terminal serine 81 phosphorylation site. *Am J Physiol Heart Circ Physiol,* 2006; 291:H1614-1622.

Beam, KG; Tanabe, T; Numa, S. Structure, function, and regulation of the skeletal muscle dihydropyridine receptor. *Ann NY Acad Sci,* 1989; 560:127–137.

Bennett, BD; Callaway, JC; Wilson, CJ. Intrinsic membrane properties underlying spontaneous tonic firing in neostriatal cholinergic interneurons. *J Neurosci,* 2000; 20:8493-8503.

Bennett, DJ; Gorassini, M; Fouad, K; Sanelli, L; Han, Y; Cheng, J. Spasticity in rats with sacral spinal cord injury. *J Neurotrauma,* 1999; 16:69-84.

Bennett, DJ; Li, Y; Siu, M. Plateau potentials in sacrocaudal motoneurons of chronic spinal rats recorded in vitro. *J Neurophysiol,* 2001; 86:1955-1971.

Bennett, DJ; Sanelli, L; Cooke, CL; Harvey, PJ; Gorassini, MA. Spastic long-lasting reflexes in the awake rat after sacral spinal cord injury. *J Neurophysiol,* 2004; 91:2247-2258.

Brandt, A; Striessnig, J; Moser, T. CaV1.3 channels are essential for development and presynaptic activity of cochlear inner hair cells. *J Neurosci,* 2003; 23:10832-10840.

Brownstone, RM. Beginning at the end: repetitive firing properties in the final common pathway. *Prog Neurobiol,* 2006; 78:156-72

Budde, T; Munsch, T; Pape, HC. Distribution of L-type calcium channels in rat thalamic neurones. *Eur J Neurosci,* 1998; 10:586-597.

Bui, TV; Ter-Mikaelian, M; Bedrossian, D; Rose, PK. Computational estimation of the distribution of L-type Ca(2+) channels in motoneurons based on variable threshold of activation of persistent inward currents. *J Neurophysiol,* 2006; 95:225-241.

Burns, BD. The mammalian cerebral cortex. London: Edward Arnold; 1958.

Burstein, R; Cliffer, KD; Giesler, GJ Jr. Cells of origin of the spinohypothalamic tract in the rat. *J Comp Neurol,* 1990; 291:329-344.

Button, DC; Gardiner, K; Marqueste, T; Gardiner, PF. Frequency-current relationships of rat hindlimb alpha-motoneurones. *J Physiol,* 2006; 573:663-677.

Carlin, KP; Jiang, Z; Brownstone, RM. Characterization of calcium currents in functionally mature mouse spinal motoneurons. *Eur J Neurosci,* 2000a; 12:1624-1634.

Carlin, KP; Jones, KE; Jiang, Z; Jordan, LM; Brownstone, RM. Dendritic L-type calcium currents in mouse spinal motoneurons: implications for bistability. *Eur J Neurosci,* 2000b; 12:1635-1646.

Catterall, WA; Perez-Reyes, E; Snutch, TP; Striessnig, J. International Union of Pharmacology. XLVIII. Nomenclature and structure-function relationships of voltage-gated calcium channels. *Pharmacol Rev,* 2005; 57:411-425.

Charles, AC; Piros, ET; Evans, CJ; Hales, TG. L-type Ca2+ channels and K+ channels specifically modulate the frequency and amplitude of spontaneous Ca2+ oscillations and have distinct roles in prolactin release in GH3 cells. *J Biol Chem,* 1999; 274: 7508–7515.

Chung, YH; Shin, C; Park, KH; Cha, CI. Immunohistochemical study on the distribution of neuronal voltage-gated calcium channels in the rat cerebellum. *Brain Res,* 2000; 865:278-282.

Clark, NC; Nagano, N; Kuenzi, FM; Jarolimek, W; Huber, I; Walter, D; Wietzorrek, G; Boyce, S; Kullmann, DM; Striessnig, J; Seabrook, GR. Neurological phenotype and synaptic function in mice lacking the CaV1.3 alpha subunit of neuronal L-type voltage-dependent Ca2+ channels. *Neuroscience,* 2003; 120:435-442.

Comunanza, V; Marcantoni, A; Vandael, DH; Mahapatra, S; Gavello, D; Carabelli ,V; Carbone, E. CaV1.3 as pacemaker channels in adrenal chromaffin cells: specific role on exo- and endocytosis? *Channels (Austin),* 2010; 4:440-446.

Cramer, NP; Li, Y; Keller, A. The whisking rhythm generator: anoval mammalian network for the generation of movement. *J Neurophysiol,* 2007; 97:2148-2158.

Crill, WE. Persistent sodium current in mammalian central neurons. *Annu Rev Physiol,* 1996; 58:349-362.

Dai, S; Hall, DD; Hell, JW. Supramolecular assemblies and localized regulation of voltage-gated ion channels. *Physiol Rev,* 2009; 89:411-452.

de Oliveira, RB; Howlett, MC; Gravina, FS; Imtiaz, MS; Callister, RJ; Brichta, AM; van Helden, DF. Pacemaker currents in mouse locus coeruleus neurons. *Neuroscience,* 2010; 170:166-77.

Delgado-Lezama, R; Perrier, JF; Hounsgaard, J. Local facilitation of plateau potentials in dendrites of turtle motoneurones by synaptic activation of metabotropic receptors. *J Physiol,* 1999, 515:203-207.

Delgado-Lezama, R; Perrier, JF; Nedergaard, S; Svirskis, G; Hounsgaard, J. Metabotropic synaptic regulation of intrinsic response properties of turtle spinal motoneurones. *J Physiol,* 1997; 504:97-102.

Dobremez, E; Bouali-Benazzouz, R; Fossat, P; Monteils, L; Dulluc, J; Nagy, F; Landry, M. Distribution and regulation of L-type calcium channels in deep dorsal horn neurons after sciatic nerve injury in rats. *Eur J Neurosci,* 2005; 21:3321-3333.

Dolphin, AC. G protein modulation of voltage-gated calcium channels. *Pharmacol Rev,* 2003; 55:607-627.

Dunlap, K; Luebke, JI; Turner, TJ. Exocytotic Ca2-channels in mammalian central neurons. *Trends Neurosci,* 1995; 18: 89–98.

Eccles, JC. The Physiology of Nerve Cells. Baltimore: The Johns Hopkins Press; 1957.

Eken, T; Hultborn, H; Kiehn, O. Possible functions of transmitter-controlled plateau potentials in alpha motoneurones. *Prog Brain Res,* 1989; 80:257-267.

Eken, T; Kiehn, O. Bistable firing properties of soleus motor units in unrestrained rats. *Acta Physiol Scand,* 1989; 136:383-394.

Elbasiouny, SM; Bennett, DJ; Mushahwar, VK. Simulation of dendritic CaV1.3 channels in cat lumbar motoneurons: spatial distribution. *J Neurophysiol,* 2005, 94:3961-3974.

ElBasiouny, SM; Schuster, JE; Heckman, CJ. Persistent inward currents in spinal motoneurons: important for normal function but potentially harmful after spinal cord injury and in amyotrophic lateral sclerosis. *Clin Neurophysiol,* 2010; 121:1669-1679.

Ertel, EA; Campbell, KP; Harpold, MM; Hofmann, F; Mori, Y; Perez-Reyes, E; Schwartz, A; Snutch, TP; Tanabe, T; Birnbaumer, L; Tsien, RW; Catterall, WA. Nomenclature of voltage-gated calcium channels. *Neuron,* 2000; 25:533-535.

Fan, QI; Vanderpool, KM; Chung, HS; Marsh, JD. The L-type calcium channel alpha 1C subunit gene undergoes extensive, uncoordinated alternative splicing. *Mol Cell Biochem,* 2005; 269:153-163.

Felten, DL; Sladek, JR Jr. Monoamine distribution in primate brain V. Monoaminergic nuclei: anatomy, pathways and local organization. *Brain Res Bull,* 1983; 10:171-284.

Forbes, A. The interpretation of spinal reflexes in terms of present knowledge of nerve conduction. *Physiol Rev,* 1922; 2:361-414.

Franzini-Armstrong, C; Protasi, F. Ryanodine receptors of striated muscles: a complex channel capable of multiple interactions. *Physiol Rev,* 1997; 77: 699–729.

Gabrielaitis, M; Buisas, R; Guzulaitis, R; Svirskis, G; Alaburda, A. Persistent sodium current decreases transient gain in turtle motoneurons. *Brain Res,* 2011; 1373:11-16.

Gartside, SE; Cole, AJ; Williams, AP; McQuade, R; Judge, SJ. AMPA and NMDA receptor regulation of firing activity in 5-HT neurons of the dorsal and median raphe nuclei. *Eur J Neurosci,* 2007; 25:3001-3008.

Giesler, GJ Jr; Björkeland, M; Xu, Q; Grant, G. Organization of the spinocervicothalamic pathway in the rat. *J Comp Neurol,* 1988; 268:223-233.

Giraldez, T; de la Pena, P; Gomez-Varela, D; Barros, F. Correlation between electrical activity and intracellular Ca^{2+} oscillations in GH3 rat anterior pituitary cells. *Cell Calcium,* 2002; 31: 65–78.

Goldman-Rakic, PS. Cellular basis of working memory. *Neuron,* 1995; 14:477-485.

Goldman-Rakic, PS. Memory: recording experience in cells and circuits: diversity in memory research. *Proc Natl Acad Sci U S A,* 1996; 93:13435-13437.

Gorassini, M; Yang, JF; Siu, M; Bennett, DJ. Intrinsic activation of human motoneurons: reduction of motor unit recruitment thresholds by repeated contractions. *J Neurophysiol,* 2002a; 87:1859-1866.

Gorassini, M; Yang, JF; Siu, M; Bennett. Intrinsic activation of human motoneurons: possible contribution to motor unit excitation. *J Neurophysiol,* 2002b; 87:1850-1858.

Grande, G; Bui, TV; Rose, PK. Estimates of the location of L-type Ca2+ channels in motoneurons of different sizes: a computational study. *J Neurophysiol,* 2007; 97:4023-4035.

Granit, R. Mechanisms regulating the discharge of motoneurons. Sherrington lectures XI. Liverpool: Liverpool University Press; 1972; pp 1-78.

Grunnet, M; Kaufmann, WA. Coassembly of big conductance Ca2+-activated K+ channels and L-type voltage-gated Ca2+ channels in rat brain. *J Biol Chem,* 2004; 279:36445-36453.

Guertin, PA; Hounsgaard, J. Non-volatile general anaesthetics reduce spinal activity by suppressing plateau potentials. *Neuroscience,* 1999; 88:353-358.

Gutman, AM. Bistability of dendrites. *Int J Neural Sys,* 1991; 1:291–304.

Hainsworth, AH; Röper, J; Kapoor, R; Ashcroft, FM. Identification and electrophysiology of isolated pars compacta neurons from guinea-pig substantia nigra. *Neuroscience,* 1991; 43:81-93.

Hajós, M; Gartside, SE; Villa, AE; Sharp, T. Evidence for a repetitive (burst) firing pattern in a sub-population of 5-hydroxytryptamine neurons in the dorsal and median raphe nuclei of the rat. *Neuroscience,* 1995; 69:189-197.

Halling, DB; Aracena-Parks, P; Hamilton, SL. Regulation of voltage-gated Ca2+ channels by calmodulin. *Sci STKE,* 2005; 315:re15.

Hanson, JE; Smith, Y. Subcellular distribution of high-voltage-activated calcium channel subtypes in rat globus pallidus neurons. *J Comp Neurol,* 2002; 442:89-98.

Harvey, PJ: Li, Y; Li, X; Bennett, DJ. Persistent sodium currents and repetitive firing in motoneurons of the sacrocaudal spinal cord of adult rats. *J Neurophysiol,* 2006; 96:1141-1157.

Hatano, S; Yamashita, T; Sekiguchi, A; Iwasaki, Y; Nakazawa, K; Sagara, K; Iinuma, H; Aizawa, T; Fu, LT. Molecular and electrophysiological differences in the L-type Ca2+ channel of the atrium and ventricle of rat hearts. *Circ J,* 2006; 70:610-614.

Heckman, CJ; Lee, RH; Brownstone, RM. Hyperexcitable dendrites in motoneurons and their neuromodulatory control during motor behavior. *Trends Neurosci,* 2003; 26:688-695.

Heckman, CJ; Gorassini, MA; Bennett, DJ. Persistent inward currents in motoneuron dendrites: implications for motor output. *Muscle Nerve,* 2005; 31:135-156.

Heckman, CJ; Mottram, C; Quinlan, K; Theiss, R; Schuster, J. Motoneuron excitability: the importance of neuromodulatory inputs. *Clin Neurophysiol,* 2009; 120:2040-2054.

Hell, JW; Westenbroek, RE; Warner, C; Ahlijanian, MK; Prystay, W; Gilbert, MM; Snutch, TP; Catterall, WA. Identification and differential subcellular localization of the neuronal class C and class D L-type calcium channel α_1 subunits. *J Cell Biol,* 1993; 123:949-962.

Helton, TD; Xu, W; Lipscombe, D. Neuronal L-type calcium channels open quickly and are inhibited slowly. *J Neurosci,* 2005; 25:10247-10251.

Hetzenauer, A; Sinnegger-Brauns, MJ; Striessnig, J; Singewald, N. Brain activation pattern induced by stimulation of L-type Ca2+-channels: contribution of Ca(V)1.3 and Ca(V)1.2 isoforms. *Neuroscience,* 2006; 139:1005-1015.

Hökfelt, T; Johansson, O; Goldstein, M. Chemical anatomy of the brain. *Science,* 1984; 25:1326-1334.

Hornby, TG; McDonagh, JC; Reinking, RM; Stuart, DG. Effects of excitatory modulation on intrinsic properties of turtle motoneurons. *J Neurophysiol,* 2002; 88:86-97.

Hounsgaard, J; Hultborn, H; Jespersen, B; Kiehn, O. Intrinsic membrane properties causing a bistable behaviour of alpha-motoneurones. *Exp Brain Res,* 1984; 55:391-394.

Hounsgaard, J; Hultborn, H; Jespersen, B; Kiehn, O. Bistability of alpha-motoneurones in the decerebrate cat and in the acute spinal cat after intravenous 5-hydroxytryptophan. *J Physiol,* 1988; 405:345-367.

Hounsgaard, J Kiehn, O. Ca++ dependent bistability induced by serotonin in spinal motoneurons. *Exp Brain Res,* 1985; 57:422-425.

Hounsgaard, J; Kiehn, O. Serotonin-induced bistability of turtle motoneurones caused by a nifedipine-sensitive calcium plateau potential. *J Physiol,* 1989; 414:265-282.

Hounsgaard, J; Kiehn, O. Calcium spikes and calcium plateaux evoked by differential polarization in dendrites of turtle motoneurones in vitro. *J Physiol,* 1993; 468:245–259.

Hounsgaard, J; Kjaerulff, O. Ca2+-Mediated Plateau Potentials in a Subpopulation of Interneurons in the Ventral Horn of the Turtle Spinal Cord. *Eur J Neurosci,* 1992; 4:183-188.

Hounsgaard, J; Mintz, I. Calcium conductance and firing properties of spinal motoneurones in the turtle. *J Physiol,* 1988; 398:591-603.

Hsiao, CF; Del Negro, CA; Trueblood, PR; Chandler, SH. Ionic basis for serotonin-induced bistable membrane properties in guinea pig trigeminal motoneurons. *J Neurophysiol,* 1998; 79:2847-2856.

Hui, A; Ellinor, PT; Krizanova, O; Wang, JJ; Diebold, RJ; Schwartz, A. Molecular cloning of multiple subtypes of a novel rat brain isoform of the alpha 1 subunit of the voltage-dependent calcium channel. *Neuron,* 1991, 7:35–44.

Hultborn, H; Brownstone, RB; Toth, TI; Gossard, J-P. Key mechanisms for setting the input-output gain across the motoneuron pool. *Prog Brain Res,* 2004; 143:77-95.

Hultborn, H; Enríquez Denton, M; Wienecke, J; Nielsen, JB. Variable amplification of synaptic excitation in spinal motoneurones by dendritic PIC. *J Physiol,* 2003; 552:945-52.

Hultborn, H; Fedrichuk, B. 2009 Spinal Motor Neurons: Properties. In: Squire LR, editor. *Encyclopedia of Neuroscience (Vol 9).* Oxford: Academic Press; 2009; pp 309-319.

Ihara, Y; Yamada. Y; Fujii, Y; Gonoi, T; Yano, H; Yasuda, K; Inagaki, N; Seino, Y; Seino, S. Molecular diversity and functional characterization of voltage-dependent calcium channels (CACN4) expressed in pancreatic beta-cells. *Mol Endocrinol,* 1995, 9:121-130.

Jackson, AC; Yao, GL; Bean, BP. Mechanism of spontaneous firing in dorsomedial suprachiasmatic nucleus neurons. *J Neurosci,* 2004; 24:7985–7998.

Jacobo, SM; Guerra, ML; Hockerman, GH. Cav1.2 and Cav1.3 are differentially coupled to glucagon-like peptide-1 potentiation of glucose-stimulated insulin secretion in the pancreatic beta-cell line INS-1. *J Pharmacol Exp Ther,* 2009; 331:724-32.

Jiang, Z; Rempel, J; Li, J; Sawchuk, MA; Carlin, KP; Brownstone, RM. Development of L-type calcium channels and a nifedipine-sensitive motor activity in the postnatal mouse spinal cord. *Eur J Neurosci,* 1999; 11:3481-3487.

Joux, N; Chevaleyre, V; Alonso, G; Boissin-Agasse, L: Moos, FC. Desarménien MG, Hussy N. High voltage-activated Ca2+ currents in rat supraoptic neurones: biophysical properties and expression of the various channel alpha1 subunits. *Neuroendocrinol,* 2001; 13:638-649.

Kang, Y; Kitai, ST. A whole cell patch-clamp study on the pacemaker potential in dopaminergic neurons of rat substantia nigra compacta. *Neurosci Res,* 1993; 18:209-221.

Koschak, A; Reimer, D; Huber, I; Grabner, M; Glossmann, H; Engel, J; Striessnig, J. Alpha 1D (Cav1.3) subunits can form l-type Ca^{2+} channels activating at negative voltages. *J Biol Chem,* 2001; 276:22100–22106.

Koulakov, AA; Raghavachari, S; Kepecs, A; Lisman, JE. Model for a robust neural integrator. *Nat Neurosci,* 2002; 5:775-782.

Lacinová, L. Voltage-dependent calcium channels. *Gen Physiol Biophys,* 2005; 24 (Suppl 1):1-78.

Leitch, B; Szostek, A; Lin, R; Shevtsova, O. Subcellular distribution of L-type calcium channel subtypes in rat hippocampal neurons. *Neuroscience,* 2009; 164:641-657.

Li, Y; Bennett, DJ. Persistent sodium and calcium currents cause plateau potentials in motoneurons of chronic spinal rats. *J Neurophysiol,* 2003; 90:857-869.

Li, Y; Gorassini, MA; Bennett, DJ. Role of persistent sodium and calcium currents in motoneuron firing and spasticity in chronic spinal rats. *J Neurophysiol,* 2004a; 91:767-783.

Li, Y; Harvey, PJ; Li, X; Bennett, DJ. Spastic long-lasting reflexes of the chronic spinal rat studied in vitro. *J Neurophysiol,* 2004b; 91:2236-2246.

Li, YQ; Takada, M; Mizuno, N. The sites of origin of serotoninergic afferent fibers in the trigeminal motor, facial, and hypoglossal nuclei in the rat. *Neurosci Res,* 1993; 17:307-313.

Liao, P; Yong, TF; Liang, MC; Yue, DT; Soong, TW. Splicing for alternative structures of Cav1.2 Ca2+ channels in cardiac and smooth muscles. *Cardiovasc Res,* 2005; 68:197-203.

Liljelund, P; Netzeband, JG; Gruol, DL. L-type calcium channels mediate calcium oscillations in early postnatal Purkinje neurons. *J Neurosci,* 2000; 20:7394–7403.

Lipscombe, D; Helton, TD; Xu, W. L-type calcium channels: the low down. *J Neurophysiol,* 2004; 92:2633-2641.

Lipscombe, D; Pan, QJ; Gray, AC. Functional diversity in neuronal voltage-gated calcium channels by alternative splicing of Cav alpah1. *Mol Neurobiol,* 2002, 26: 21–44.

Llinás, RR; Grace, AA; Yarom, Y. In vitro neurons in mammalian cortical layer 4 exhibit intrinsic oscillatory activity in the 10- to 50-Hz frequency range. *Proc Natl Acad Sci U S A,* 1991; 88:897-901.

Lo, FS; Erzurumlu, RS. L-type calcium channel-mediated plateau potentials in barrelette cells during structural plasticity. *J Neurophysiol,* 2002; 88:794-801.

Lo, FS; Guido, W; Erzurumlu, RS. Electrophysiological properties and synaptic responses of cells in the trigeminal principal sensory nucleus of postnatal rats. *J Neurophysiol,* 1999; 82:2765-2775.

Lorente de No, R. Analysis of the activity of the chains of internuncial neurons. *J Neurophysiol,* 1938; 1:207-244.

Ludwig, A; Flockerzi, V; Hofmann, F. Regional expression and cellular localization of the alpha1 and beta subunit of high voltage-activated calcium channels in rat brain. *J Neurosci,* 1997; 17:1339-1349.

Mangoni, ME; Couette, B; Bourinet, E; Platzer, J; Reimer, D; Striessnig, J; Nargeot, J. Functional role of L-type Cav1.3 Ca2+ channels in cardiac pacemaker activity. *Proc Natl Acad Sci U S A*, 2003; 100:5543-5548.

Marcantoni, A; Vandael, DH; Mahapatra, S; Carabelli, V; Sinnegger-Brauns, MJ; Striessnig, J; Carbone, E. Loss of Cav1.3 channels reveals the critical role of L-type and BK channel coupling in pacemaking mouse adrenal chromaffin cells. *J Neurosci*, 2010; 30:491-504.

Matsushita, M; Hosoya, Y. Spinocerebellar projections to lobules III to V of the anterior lobe in the cat, as studied by retrograde transport of horseradish peroxidase. *J Comp Neurol*, 1982; 208:127-143.

McCormick, DA. Neuronal networks: flip-flops in the brain. *Curr Biol*, 2005; 15:R294-296.

McKinney, BC; Sze, W; Lee, B; Murphy, GG. Impaired long-term potentiation and enhanced neuronal excitability in the amygdala of Ca(V)1.3 knockout mice. *Neurobiol Learn Mem*, 2009; 92:519-528.

Mercuri, NB; Bonci, A; Calabresi, P; Stratta, F; Stefani, A; Bernardi, G. Effects of dihydropyridine calcium antagonists on rat midbrain dopaminergic neurones. *Br J Pharmacol*, 1994; 113:831-838.

Morgans, CW. Calcium channel heterogeneity among cone photoreceptors in the tree shrew retina. *Eur J Neurosci*, 1999, 11:2989-2993.

Morisset, V; Nagy, F. Modulation of regenerative membrane properties by stimulation of metabotropic glutamate receptors in rat deep dorsal horn neurons. *J Neurophysiol*, 1996; 76:2794-2798.

Morisset, V; Nagy, F. Nociceptive integration in the rat spinal cord: role of non-linear membrane properties of deep dorsal horn neurons. *Eur J Neurosci*, 1998; 10:3642-3652.

Morisset, V; Nagy, F. Ionic basis for plateau potentials in deep dorsal horn neurons of the rat spinal cord. *J Neurosci*, 1999; 19:7309-7316.

Moritz, AT; Newkirk, G; Powers, RK; Binder, MD. Facilitation of somatic calcium channels can evoke prolonged tail currents in rat hypoglossal motoneurons. *J Neurophysiol*, 2007; 98:1042-1047.

Namkung, Y; Skrypnyk, N; Jeong, MJ; Lee, T; Lee, MS; Kim, HL; Chin, H; Suh, PG; Kim, SS; Shin, HS. Requirement for the L-type Ca(2+) channel alpha(1D) subunit in postnatal pancreatic beta cell generation. *J Clin Invest*, 2001; 108:1015-1022.

Nedergaard, S; Flatman, JA; Engberg, I. Nifedipine- and omega-conotoxin-sensitive Ca2+ conductances in guinea-pig substantia nigra pars compacta neurones. *J Physiol*, 1993; 466:727-747.

Olson, PA; Tkatch, T; Hernandez-Lopez, S; Ulrich, S; Ilijic, E; Mugnaini, E; Zhang, H; Bezprozvanny, I; Surmeier, DJ. G-Protein-Coupled Receptor Modulation of Striatal CaV1.3 L-Type Ca2+ Channels Is Dependent on a Shank-Binding Domain. *J Neurosci*, 2005; 25:1050-1062.

Ouardouz, M; Nikolaeva, MA; Coderre, E; Zamponi, GW; McRory, JE; Trapp, BD; Yin, X; Wang, W; Woulfe, J; Stys, PK. Depolarization-induced Ca^{2+} release in ischemic spinal cord white matter involves L-type Ca^{2+} channel activation of ryanodine receptors. *Neuron,* 2003; 40:53–63.

Paroschy, KL; Shefchyk, SJ. Non-linear membrane properties of sacral sphincter motoneurones in the decerebrate cat. *J Physiol,* 2000; 523:741-753.

Paxinos, G; Watson, C. The rat brain: in stereotaxic coordinates. 6th edi. London, Amsterdam, Burlinton: Academic Press; 2007.

Pennartz, CM; de Jeu, MT; Bos, NP; Schaap, J; Geurtsen, AM. Diurnal modulation of pacemaker potentials and calcium current in the mammalian circadian clock. *Nature,* 2002; 416:286–290.

Perez-Reyes, E; Wei, XY; Castellano, A; Birnbaumer, L. Molecular diversity of Ltype calcium channels. Evidence for alternative splicing of the transcripts of three non-allelic genes. *J Bio Chem,* 1990; 265:20430–20436.

Perrier, JF; Alaburda, A; Hounsgaard, J. Spinal plasticity mediated by postsynaptic L-type Ca2+ channels. *Brain Res Brain Res Rev,* 2002; 40:223-229.

Perrier, JF; Mejia-Gervacio, S; Hounsgaard, J. Facilitation of plateau potentials in turtle motoneurones by a pathway dependent on calcium and calmodulin. *J Physiol,* 2000; 528:107-113.

Peterson, BZ; DeMaria, CD; Adelman, JP; Yue, DT. Calmodulin is the Ca2+ sensor for Ca2+ -dependent inactivation of L-type calcium channels. *Neuron,* 1999; 22:549-558.

Placantonakis, D; Welsh, J. Two distinct oscillatory states determined by the NMDA receptor in rat inferior olive. *J Physiol,* 2001; 534:123–140.

Platzer, J; Engel, J; Schrott-Fischer, A; Stephan, K; Bova, S; Chen, H; Zheng, H, Striessnig, J. Congenital deafness and sinoatrial node dysfunction in mice lacking class D L-type Ca2+ channels. *Cell,* 2000; 102:89-97.

Puopolo, M; Raviola, E; Bean, BP. Roles of subthreshold calcium current and sodium current in spontaneous firing of mouse midbrain dopamine neurons. *J Neurosci,* 2007; 27:645-656.

Powers, RK; Binder, MD. Input-output functions of mammalian motoneurons. *Rev Physiol Biochem Pharmacol,* 2001; 143:137-263.

Powers, RK; Binder, MD. Persistent sodium and calcium currents in rat hypoglossal motoneurons. *J Neurophysiol,* 2003; 89:615-624.

Puopolo, M; Raviola, E; Bean, BP. Roles of subthreshold calcium current and sodium current in spontaneous firing of mouse midbrain dopamine neurons. *J Neurosci,* 2007; 27:645-656.

Putzier, I; Kullmann, PH; Horn, JP; Levitan, ES. Cav1.3 channel voltage dependence, not Ca2+ selectivity, drives pacemaker activity and amplifies bursts in nigral dopamine neurons. *J Neurosci,* 2009; 29:15414-15419.

Qu, Y; Baroudi, G; Yue, Y; El-Sherif, N; Boutjdir, M. Localization and modulation of {alpha}1D (Cav1.3) L-type Ca channel by protein kinase A. *Am J Physiol Heart Circ Physiol,* 2005; 288:H2123-2130.

Rall, W. Path to biophysical insights about dendrites and synaptic function. In: Samson F and Adelman G, editors. *The Neurosciences: Paths of Discovery (Vol. II)*. Boston: Birkhäuser; 1992; pp 214–238.

Reuter, H. A variety of calcium channels. *Nature*, 1985; 316: 391.

Rhodes, KJ; Trimmer, JS. Antibodies as valuable neuroscience research tools versus reagents of mass distraction. *J Neurosci*, 2006; 26:8017-8020.

Russo, RE; Hounsgaard, J. Plateau-generating neurones in the dorsal horn in an in vitro preparation of the turtle spinal cord. *J Physiol*, 1996; 493:39-54.

Russo, RE; Nagy, F; Hounsgaard, J. Modulation of plateau properties in dorsal horn neurones in a slice preparation of the turtle spinal cord. *J Physiol*, 1997; 499:459-474.

Safa, P; Boulter, J; Hales, TG. Functional properties of $Ca_V1.3$ (alpha1D) L-type Ca2+ channel splice variants expressed by rat brain and neuroendocrine GH3 cells. *J Biol Chem*, 2001; 276:38727-38737.

Sanchez-Vives, MV; McCormick, DA. Cellular and network mechanisms of rhythmic recurrent activity in neocortex. *Nat Neurosci*, 2000; 3:1027-1034.

Sand, O; Chen, BM; Grinnell, AD. Contribution of L-type Ca(2+) channels to evoked transmitter release in cultured *Xenopus* nerve-muscle synapses. *J Physiol*, 2001; 536:21–33.

Sandler, VM; Puil, E; Schwarz, DW. (1998) Intrinsic response properties of bursting neurons in the nucleus principalis trigemini of the gerbil. *Neuroscience*, 1998; 83:891-904.

Saper, CB. An open letter to our readers on the use of antibodies. *J Comp Neurol*, 2005; 493:477-478.

Saper, CB. A guide to the perplexed on the specificity of antibodies. *J Histochem Cytochem*, 2009; 57:1-5.

Schneider, MF; Chandler, WK. Voltage dependent charge movement of skeletal muscle: a possible step in excitation-contraction coupling. *Nature*, 1973, 242:244–246.

Scholze, A; Plant, TD; Dolphin, AC; Nurnberg, B. Functional expression and characterization of a voltage-gated CaV1.3 (alpha1D) calcium channel subunit from an insulin-secreting cell line. *Mol Endocrinol*, 2001; 15:1211–1221.

Schwindt, PC; Crill, WE. A persistent negative resistance in cat lumbar motoneurons. *Brain Res*, 1977; 120:173-178.

Schwindt, PC; Crill, WE. Properties of a persistent inward current in normal and TEA-injected motoneurons. *J Neurophysiol*, 1980; 43:1700-1724.

Schwindt, PC; Crill, WE. Amplification of synaptic current by persistent sodium conductance in apical dendrite of neocortical neurons. *J Neurophysiol*, 1995; 74:2220-2224.

Simon, M; Perrier, JF; Hounsgaard, J. Subcellular distribution of L-type Ca2+ channels responsible for plateau potentials in motoneurons from the lumbar spinal cord of the turtle. *Eur J Neurosci*, 2003; 18:258-266.

Singh, A; Gebhart, M; Fritsch, R; Sinnegger-Brauns, MJ; Poggiani, C; Hoda, JC; Engel, J; Romanin, C; Striessnig, J; Koschak, A. Modulation of voltage- and Ca2+-dependent gating of CaV1.3 L-type calcium channels by alternative splicing of a C-terminal regulatory domain. *J Biol Chem*, 2008; 283:20733-20744.

Sinnegger-Brauns, MJ; Hetzenauer, A; Huber, IG; Renström, E; Wietzorrek, G; Berjukov, S; Cavalli, M; Walter, D; Koschak, A; Waldschütz, R; Hering, S; Bova, S; Rorsman, P; Pongs, O; Singewald, N; Striessnig, J. soform-specific regulation of mood behavior and pancreatic beta cell and cardiovascular function by L-type Ca 2+ channels. *J Clin Invest,* 2004; 113:1430-1439.

Smith, M; Perrier, JF. Intrinsic properties shape the firing pattern of ventral horn interneurons from the spinal cord of the adult turtle. *J Neurophysiol,* 2006; 96:2670-2677.

Smith, PA; Aschroft, FM; and Fewtrell, CM. Permeation and gating properties of the L-type calcium channel in mouse pancreatic beta cells. *J Gen Physiol,* 1993; 101:767–797

Snutch, TP; Tomlinson, WJ; Leonard, JP; Gilbert, MM. Distinct calcium channels are generated by alternative splicing and are differentially expressed in the mammalian CNS. *Neuron,* 1991; 7:45–57.

Storm, JF; Vervaele, K; Hu, H; Graham, LJ. Functions of the persitent Na+ current in cortical neurons revealed by dynamic clamp. In: Destexhe A and Bal T, editors. *Dynamic Clamp: From Principles to Applications.* Springer Press; 2009; pp 165-197.

Striessnig, J; Hoda, JC; Koschak, A; Zaghetto, F; Müllner, C; Sinnegger-Brauns, MJ; Wild, C; Watschinger, K; Trockenbacher, A; Pelster, G. L-type Ca2+ channels in Ca2+ channelopathies. *Biochem Biophys Res Commun,* 2004; 322:1341-1346.

Striessnig, J; Koschak, A; Sinnegger-Brauns, MJ; Hetzenauer, A; Nguyen, NK; Busquet, P; Pelster, G; Singewald, N. Role of voltage-gated L-type Ca2+ channel isoforms for brain function. *Biochem Soc Trans,* 2006; 34:903-909.

Striessnig, J; Bolz, HJ; Koschak, A. Channelopathies in Cav1.1, Cav1.3, and Cav1.4 voltage-gated L-type Ca2+ channels. *Pflugers Arch,* 2010; 460:361-374.

Sukiasyan, N; Hultborn, H; Zhang, M. Distribution of calcium channel Ca(V)1.3 immunoreactivity in the rat spinal cord and brain stem. *Neuroscience,* 2009; 159:217-235.

Surmeier, DJ; Seno, N; Kitai, ST. Acutely isolated neurons of the rat globus pallidus exhibit four types of high-voltage-activated Ca2+ current. *J Neurophysiol,* 1994; 71:1272-1280.

Svirskis, G; Gutman, A; Hounsgaard, J. Electrotonic structure of motoneurons in the spinal cord of the turtle: inferences for the mechanisms of bistability. *J Neurophysiol,* 2001; 85:391-398.

Svirskis, G; Hounsgaard, J. Transmitter regulation of plateau properties in turtle motoneurons. *J Neurophysiol,* 1998; 79:45-50.

Tachibana, M; Okada, T; Arimura, T; Kobayashi, K. Dihydropyridine-sensitive calcium current mediates neurotransmitter release from retinal bipolar cells. *Ann NY Acad Sci,* 1993; 707:359–361.

Takada, M; Kang, Y; Imanishi, M. Immunohistochemical localization of voltage-gated calcium channels in substantia nigra dopamine neurons. *Eur J Neurosci,* 2001; 13:757-762.

Takimoto, K; Li, D; Nerbonne, JM; Levitan, ES. Distribution, splicing and glucocorticoid-induced expression of cardiac alpha 1C and alpha ID voltage-gated Ca2+ channel mRNAs. *J Mol Cell Cardiol,* 1997; 29:3035–3042.

Tanabe, T; Mikami, A; Numa, S; Beam, KG. Cardiac-type excitation-contraction coupling in dysgenic skeletal muscle injected with cardiac dihydropyridine receptor cDNA. *Nature,* 1990; 344:451–453.

Tanaka, O; Sakagami, H; Kondo, H. Localization of mRNAs of voltage-dependent Ca(2+)-channels: four subtypes of alpha 1- and beta-subunits in developing and mature rat brain. *Brain Res Mol Brain Res,* 1995; 30:1-16.

Theiss, RD; Kuo, JJ; Heckman, CJ. Persistent inward currents in rat ventral horn neurones. *J Physiol,* 2007; 580:507-522.

Timmermann, DB; Westenbroek, RE; Schousboe, A; Catterall, WA. Distribution of high-voltage-activated calcium channels in cultured gamma-aminobutyric acidergic neurons from mouse cerebral cortex. *J Neurosci Res,* 2002; 67:48-61.

Tsukahara, N; Bando, T; Kitai, ST; Kiyohara, T. Cerebello-pontine reverbearating circuit. *Brain Res,* 1971; 33:233-237.

Tsukahara, N; Bando, T; Murakami, F; Oda, Y. Properties of cerebello-precerebellar reverberating circuits. *Brain Res,* 1983; 274:249-259.

Ungerstedt, U. Stereotaxic mapping of the monoamine pathways in the rat brain. *Acta Physiol Scand,* 1971; 2 (Suppl 367):1–48.

Vandael, DH; Marcantoni, A; Mahapatra, S; Caro, A; Ruth, P; Zuccotti, A; Knipper, M; Carbone, E. Ca(v)1.3 and BK channels for timing and regulating cell firing. *Mol Neurobiol,* 2010; 42:185-198.

Vergara, R; Rick, C; Hernandez-Lopez, S; Laville, JA; Guzman, JN; Galarraga, E; Surmeier, DJ; Bargas, J. Spontaneous voltage oscillations in striatal projection neurons in a rat corticostriatal slice. *J Physiol,* 2003; 553:169–182.

Viard, P; Butcher, AJ; Halet, G; Davies, A; Nürnberg, B; Heblich, F; Dolphin, AC. PI3K promotes voltage-dependent calcium channel trafficking to the plasma membrane. *Nat Neurosci,* 2004; 7:939-946.

Wang, XJ. Synaptic reverberation underlying mnemonic persistent activity. *Trends Neurosci,* 2001; 24:455-463.

Weisskopf, MG; Bauer, EP; LeDoux, JE. L-type voltage-gated calcium channels mediate NMDA-independent associative long-term potentiation at thalamic input synapses to the amygdala. *J Neurosci,* 1999; 19:10512-10519.

Westenbroek, RE; Ahlijanian, MK; Catterall, WA. Clustering of L-type Ca2+ channels at the base of major dendrites in hippocampal pyramidal neurons. *Nature,* 1990; 347:281-284.

Westenbroek, RE; Hoskins, L; Catterall, WA. Localization of Ca2+ channel subtypes on rat spinal motor neurons, interneurons, and nerve terminals. *J Neurosci,* 1998; 18:6319-6330.

Wienecke, J; Westerdahl, A-C; Hultborn, H; Kiehn, O; Ryge, J. Global gene expression analysis of rodent motor neurons following spinal cord injury link molecular mechanisms to development of post-injury spasticity. *J. Neurophysiol,* 2010; 103:761-778.

Williams, JT; North, RA; Shefner, SA; Nishi, S; Egan, TM. Membrane properties of rat locus coeruleus neurones. *Neuroscience,* 1984; 13:137-156.

Williams, ME; Feldman, DH; McCue, AF; Brenner, R; Velicelebi, G; Ellis, SB; Harpold, MM. Structure and functional expression of alpha 1, alpha 2, and beta subunits of a novel human neuronal calcium channel subtype. *Neuron,* 1992; 8:71-84.

Wilson, CJ; Callaway, JC. Coupled oscillator model of the dopaminergic neuron of the substantia nigra. *J Neurophysiol,* 2000; 83:3084-3100.

Wilson, CJ; Groves, PM. Spontaneous firing patterns of identified spiny neurons in the rat neostriatum. *Brain Res,* 1981; 220:67-80.

Wilson, CJ; Kawaguchi, Y. The origins of two-state spontaneous membrane potential fluctuations of neostriatal spiny neurons. *J Neurosci,* 1996; 16:2397-2410.

Wiser, O; Trus, M; Hernandez, A; Renstrom, E; Barg, S; Rorsman, P; Atlas, D. The voltage sensitive Lc-type Ca^{2+} channel is functionally coupled to the exocytotic machinery. *Proc Natl Acad Sci USA,* 1999; 96:248–253.

Xu, W; Lipscombe, D. Neuronal Ca(V)1.3alpha(1) L-type channels activate at relatively hyperpolarized membrane potentials and are incompletely inhibited by dihydropyridines. *J Neurosci,* 2001; 21: 5944–5951.

Yang, SN; Larsson, O; Bränström, R; Bertorello, AM; Leibiger, B; Leibiger, IB; Moede, T; Köhler, M; Meister, B; Berggren, PO. Syntaxin 1 interacts with the L(D) subtype of voltage-gated Ca(2+) channels in pancreatic beta cells. *Proc Natl Acad Sci U S A,* 1999; 96:10164-10169.

Zhang, H; Maximov, A; Fu, Y; Xu, F; Tang, TS; Tkatch, T; Surmeier, DJ; Bezprozvanny, I. Association of CaV1.3 L-type calcium channels with Shank. *J Neurosci,* 2005; 25:1037-1049.

Zhang, M; Sukiasyan, N; Moller, M; Bezprozvanny, I; Zhang, H, Wienecke, J; Hultborn, H. Localization of L-type calcium channel Ca$_V$1.3 in cat lumbar spinal cord – with emphasis on motoneurons. *Neurosci Lett,* 2006; 407:42-47.

Zhang, M; Wienecke, J; Hultborn, H. Immunohistochemical changes of Cav1.3 and Nav1.3 channel alpha 1 subunits in rat sacro-caudal spinal motoneurons after acute and chronic spinal cord injury. *Soc Neurosci Abstr,* San Diego, 2007.

Zhang, M; Møller, M; Broman, J; Sukiasyan, N; Wienecke, J; Hultborn, H. Expression of calcium channel Ca$_V$1.3 in cat spinal cord: light and electron microscopic immunohistochemical study. *J Comp Neurol,* 2008a; 507:1109-1127.

Zhang, M; Wienecke, J; Møller, M; Hultborn, H. Dendritic distribution of Cav1.3 channels in spinal lumbar motoneurons of cats: a combined intracellular labeling and immunohistochemical study. *Soc Neurosci Abstr,* Washington DC, 2008b.

Zhang, Q; Timofeyev, V; Qiu, H; Lu, L; Li, N; Singapuri, A; Torado, CL; Shin, HS; Chiamvimonvat, N. Expression and roles of Cav1.3 (α1D) L-type Ca^{2+} channel in atrioventricular node automaticity. *J Mol Cell Cardiol,* 2011; 50:194-202.

Zhang, Z; Séguéla, P. Metabotropic induction of persistent activity in layers II/III of anterior cingulate cortex. *Cereb Cortex,* 2010; 20:2948-2957.

Zhang, Z; Xu, Y; Song, H; Rodriguez, J; Tuteja, D; Namkung, Y; Shin, HS; Chiamvimonvat, N. Functional roles of Ca(v)1.3 (alpha(1D)) calcium channel in sinoatrial nodes: insight gained using gene-targeted null mutant mice. *Circ Res,* 2002; 90: 981–987.

Chapter 2

THE DISTINCT SIGNALING MECHANISMS OF N-TYPE AND L - TYPE CALCIUM CHANNELS IN CARDIAC HYPERTROPHY AND FAILURE

N. Petrashevskaya[†], E. Kobrinsky and A. Schwartz
Institute of Molecular Pharmacology and Biophysics,
University of Cincinnati College of Medicine,
Department of Surgery, Ohio, U. S.

Abstract

Pathologic cardiac hypertrophy is an independent risk factor for myocardial infarction, arrhythmias, and subsequent heart failure. Hypertrophic remodeling and associated hypertrophic growth, fibrosis, cavity dilatation and electrophysiological remodeling occur in response to hemodynamic stresses such as hypertension, myocardial infarction, and valvular insufficiencies. The prevalence of hypertrophy and heart failure is increasing rapidly worldwide, and effective treatments remain elusive. Cardiac hypertrophy and failure is a complex syndrome, which involves neuro-endocrine stimulation and activation of Ca^{2+} signaling pathways predominantly within cellular microdomains to induce pathologic pro-growth signaling cascades. Developing new therapies to target pathological remodeling of the hearth should synergistically target excessive sympathetic activation and molecular mechanisms of ventricular remodeling

New complex heart failure therapies are based on N-type and L- type calcium channel blockers and Ca-dependent signaling pathways modulators that target sympathetic hyper-activation and Ca^{2+} dependent proximal signaling with intent to prevent, halt, or reverse the progression of disease.

[†] Address correspondence to: Natalia Petrashevskaya, PhD, Laboratory of Cardiovascular Sciences, National Institute of Aging, 5600 Nathan Shock Drive, Baltimore, MD 21224, Telephone: 410-558-8108, Fax: 410-558-8150, E-mail petrashevskayn@mail.nia.niah.gov

INTRODUCTION

Cardiovascular disease remains the primary cause of mortality in the Western world, with heart failure representing the fastest-growing subclass over the past decade (Hobbs R.E. et al. 2004, Levy D. et al. 2002). There are approximately 5 million Americans currently diagnosed with heart failure, which is characterized by a 5-year survival rate of approximately 50% and which has a total economic impact as high as $100 billion per year. Heart failure is a complex and progressive syndrome that often culminates in death by arrhythmias or hemodynamic instability. Heart failure is characterized by ventricular dilation, elevated systolic wall stress, dysregulation in Ca^{2+} homeostasis within the myocyte, and activation of the sympathetic nervous system (Haldeman G.et al 1999, Malek M. 1999, Lloyd-Jones D. et al. 2002). Poor ventricular pump function in heart failure activates reflex pathways that lead to increased catecholamine release within the heart and endocrine tissues in an attempt to maintain adequate perfusion (Packer M. 2001, Bristow M. 2000, Bouzamondo A. et al. 2001, Houser S.et al. 2000).

At the cellular level, cardiomyocyte hypertrophy is characterized by an increase in cell size, enhanced protein synthesis, activation of fetal genes, and cytoskeleton reorganization. A number of effectors are known to induce cardiac cell hypertrophy *in vitro*, including angiotensin II (Ang II) (Sadoshima J. et al 1995), catecholamines (Rockman H. et al. 2002), endothelin (Sudgen P. et al. 2003). Many, but not all, of these hormones are known to be coupled to alterations in Ca^{2+} homeostasis. In particular, Ang II and endothelin (Mackenzie L . et all. 2002) are coupled to Inositol 1,4,5-triphosphate (InsP3) generation and Ca^{2+} mobilization from stores. Catecholamines, in particular through β1 adrenoreceptors (β1-AR), are known to induce an increase in the frequency and amplitude of Ca^{2+} spiking in cardiomyocytes. Heart failure is one of the most debilitating diseases of the cardiovascular system and involves, at least in part, a maladaptive regulation of calcium signaling in the heart, in which normal β-adrenergic regulation of L-type calcium channels (Cav1.2) fails (Marks A. 2005). Neurohormonal antagonism remains the primary and the most successful goal for drug research in heart failure therapeutics.

In this chapter, we have addressed the role of the signaling pathways initiated by high voltage-activated calcium channels, L-type and N-type calcium channels, through which a variety of heurohormonal stresses induce cardiac hypertrophy in vitro and in vivo and discussed how this pathways can be used as a target of complex therapeutic interventions with well-known and novel drugs.

1. VOLTAGE-GATED CALCIUM CHANNELS IN THE HEART

There are two types of high voltage-activated (HVA) calcium channels that are essential for the function of the heart. L –type ($Ca_v1.2$, $Ca_v1.3$) is responsible for excitation – contraction coupling and expressed in the cardiomyocites. N-type ($Ca_v2.1$) is responsible for release of catecholamines and acetylcholine and expressed in presynaptic regions of sympathetic and parasympathetic nervous system.

The $Ca_v1.2$ channels are activated by membrane depolarization to potentials > -40 mV, undergo voltage- and Ca^{2+}-dependent inactivation, and are sensitive to dihydropyridines. The pore-forming α- subunits of these channels are transcribed from different genes. L- type calcium channel α subunits presented by two genes in the cardiac myocytes. The $Ca_v1.2$ gene (CACNA1C or α_{1C}) encodes for the typical L-type Ca2+ channel in ventricular and atrial myocytes (see for review Bodi et al.,2005), and $Ca_v1.3$ (CACNA1D or $\alpha1D$), which produces currents endowed with a more negative activation threshold and faster kinetics compared to classical L-type, is mainly expressed in atrioventricular and sinus nodes.

$Ca_v1.2$ is organized into four homologous domains (I–IV), each consisting of six-transmembrane α helices (S1 through S6) and a membrane-associated P loop between S5 and S6. There are two size forms of the $Ca_v1.2$ subunit of channels, 240 and 210 kDa, different by truncation of the C-terminus (Hulme et al. 2006, Fuller M. et al. 2010). L-type calcium channel C-terminus is proteolytically cleaved and then rebinds to the rest of the channel protein to mediate an autoinhibition. This complex is augmented by the binding of one of the A-kinase anchoring proteins, AKAP-15, which brings protein kinase A (PKA) into the close vicinity of the channel. Activation of PKA then results in the phosphorylation of a key serine residue that, in turn, results in a disruption of the interaction between the channel and the cleaved C-terminus. This then removes the autoinhibition, leading to enhanced L-type currents and, consequently, increased heart rate. This mechanism is responsible for the classical enhancement of strength of cardiac contraction by adrenaline and noradrenaline, which largely depends on beta-adrenergic receptor and PKA-mediated enhancement of $Ca_v1.2$ channel.

All these channels are multi-subunit complexes consisting of a pore forming α_1 subunit and auxillary subunits β, $\alpha2\delta$, and γ. General model proposed for α_1 subunit forms a highly Ca^{2+} selective voltage gated pore where auxiliary subunits interact with α_1 subunit and modulate its activity without effect on its selectivity.

There are four subfamilies of $Ca_v\beta s$ (β_1- β_4), each with splice variants, encoded by four distinct genes. Ca^{2+} current was dramatically increased when the α_1 and β subunits are co-expressed in heterologous expression systems. $Ca_v\beta$ subunits are important for trafficking α_1 pore subunits to the plasma membrane. It has been shown recently (Altier C. et al. 2011) that β subunits prevent the premature degradation of the channels via the ER-associated degradation pathway. This pathway involves re-translocation of the channels out of the ER, ubiquitinylation and proteasomal decay. $Ca_v\beta$ subunit also modulates kinetics of activation and inactivation of $Ca_v1.2$ or $Ca_v2.2$ channels. The existence of four different genes (β_1- β_4) and extensive differential splicing give rise to multiple isoforms. The main interaction site on α_1 for β subunits is a sequence of 18 residues in the loop between domain I and II (loop I/II) called the β interaction domain or AID, which binds to a hydrophobic grove in the GK domain of β subunit. Additional interaction sites for β subunits have been identified in the NH2- and COOH-terminal regions of different α_1 subunits (Lao et al. 2008). Structural organization of $Ca_v1.2$ channels , such as the size of the cluster and intermolecular distances between α_1 subunits depends on which isoform of β subunit serve as partner of α_1 subunit (Kobrinsky E. et al. 2009). Oligomerization of $Ca_v\beta$ subunits has effect on Ca^{2+} channel activity (Lao et al. 2010). In the heart the predominant isoform is the β_{2a} subunit, but β_3 and β_4 were also detected.

In N- type $Ca_v2.2$ channels the loop I/II of the $Ca_v2.2$ channel α_1 subunits is the major binding site for not only the auxiliary β subunits but also for the G$\beta\gamma$ dimmer. G$\beta\gamma$ dimers bind to two different segments in loop I/II. Gβγ binding results in a slowing of ion current activation kinetics, which likely mirrors the voltage-dependent release from inhibition. Strong depolarization to positive potentials release Gβγ from the channel, thereby increasing channel activity. Regulation of $Ca_v2.2$ channels by G-protein coupled receptors by interaction of Gβγ with $α_1$ subunits resulted in the reduction of the amount of calcium entering synaptic nerve terminal (Tedford T.et al. 2006).

The calcium channel $α_2δ$ subunit is important for the function and trafficking of high-voltage-activated calcium channel. A total of four genes (CACNA2D1–4) with several splice variants code for $α_2δ$, but the ubiquitously expressed α2δ-1 is the major isoform in cardiac muscle, while α2δ-3 mRNA is also detected in human hearts but not in mouse (Davies A. et al. 2007). The calcium channel $α_2δ$ subunit is the receptor for anti-epileptic and anti-nociceptive drugs gabapentin and pregabalin.

It has been shown recently (Yang L. et al. 2011) that $γ_4$, $γ_6$, $γ_7$, and $γ_8$ subunits physically interact with the $Ca_v1.2$ complex in cardiac cells. The γ subunits consist of 4 transmembrane domains and differentially modulate $Ca_v1.2$ channel function when co-expressed with the $β_{1b}$ and $α_2δ$ subunits in HEK cells, altering both activation and inactivation properties. The effects of γ subunit on $Ca_v1.2$ function are dependent on the subtype of β subunit.

All HVA calcium channels have the ability to interact with calmodulin (CaM) on their C-terminus, inducing both Ca^{2+}-dependent inactivation (CDI) and facilitation, and the binding site involves variations on an isoleucine–glutamine-containing (IQ) domain (Pitt G. 2007). David Yue lab shows (Liu X. et al. 2010) that the CaM can bind and unbind channels according to naturally occurring CaM concentration fluctuations in real cells. The C-terminal region of neuronal $Ca_v1.3$ L-type calcium channels contains a functional domain that regulates the affinity of the channel for Ca^{2+}– free apo-calmodulin (apoCaM). This is an important intrinsic mechanism that allows different types of calcium channels to sense the free calmodulin concentration and respond to them by altering activity. Since CaM itself can bind tightly to both L-type ($Ca_v1.2$ and $Ca_v1.3$) and N-type ($Ca_v2.2$) Ca^{2+} channels, even in the form of apoCaM, it should be considered as a subunit. Exogenously expressed CaM can substitute for $Ca_vβ$ subunits in trafficking calcium channels (Ravindran A. et al. 2008), and is therefore playing a structural role.

2. INTEGRATION OF cAMP AND CALCIUM SIGNALING IN REGULATION OF CARDIAC CONTRACTILITY

Intracellular calcium Ca^{2+} mediates excitation-contraction (EC), excitation-transcription, excitation-secretion and a plethora of cellular responses with the high affinity and specificity that are hallmark of a regulatory second messenger (Carafoli E. et al. 2001). The primary function of the heart is to contract and pump blood in proportion to tissue needs. Each individual cardiac contraction systole depends on transient increase in the cytosolic free calcium concentration, so called "contractile calcium" to activate the contractile proteins that elicit pressure development and ejection of blood (Houser S. et al. 2008). Intracellular calcium is the key regulator of cardiac contractility and long-term alterations in calcium transients activate down-stream calcium-dependent cellular signaling pathways leading to development of cardiac hypertrophy and failure. Regulation of contractile Ca^{2+} in cardiac myocytes involves a highly specialized system of ion channels, pumps, exchangers, and microdomains (Bodi I et al. 2005). Contraction starts with depolarization of the sarcolemma that stimulates the opening of voltage-gated L-type Ca^{2+} channels with the transverse T tubules (invagination of the sarcolemma) and calcium enters a sub-membrane domains to trigger Ca^{2+} release from the sarcoplasmic reticulum (SR) via calcium-release channels, ryanodine receptors 2 (RYR2). This amplifying process, termed calcium-induced calcium release, causes a rapid increase in intracellular calcium concentration to a level required for optimal binding of Ca^{2+} to troponin C and induction of contraction. Increase in calcium concentration within microdomains required to stimulate the opening of RYR and subsequent Ca^{2+} release are greater than those in the bulk cytoplasm. Relaxation occurs when Ca^{2+} is resequestered back into the SR through the action of the SR Ca^{2+} adenosine triphosphatase (SERCA) and removed from the cells by the sarcolemmal Na^+/Ca^{2+} exchanger. Excitation-contraction coupling gain is defined as the amount of SR Ca^{2+} released for a given trigger (Ca^{2+} influx via L-type voltage –dependent calcium channels). Alteration in   EC coupling, such as a decrease in intracellular calcium transient and increased diastolic and/or systolic calcium concentration, is a major defect of the hypertrophied and dilated heart with a reduced ejection fraction. The frequency and amplitude of the systolic Ca2+ transient are regulated by neuroendocrine inputs and effectors to dynamically alter myocyte contractility and cardiac output. Numerous calcium-sensing pathways, such as calmodulin/dependent kinases and phosphatases, adenylyl cyclases (AC) and nitric oxide synthase (NOS) have evolved in cardiomyocytes to maintain calcium fluxes in a dynamic and tightly regulated balance, ensuring both adequate calcium intracellular signaling and contractile performance (for review Petrashevskaya N. et al 2002, Grimm M. et al. 2010). It is not surprising that the calcium-regulating systems are subject to regulatory influences from the autonomic nervous system (ANS), transforming information into rapid physiological responses. It is well established that the ANS is a major regulator of cardiac function, and the cardiovascular system in general. The result of this regulation is a broad variation in patterns of calcium cycling, which allows the heart to function effectively during different environmental challenges. Myocardial β-adrenoreceptors (β-AR) mediate for example, increases (the "fight or flight reaction") in the heart rate (3 fold), the force of heart contractility (5 fold) and the rate of cardiac relaxation in response to changes in sympathetic input. Alterations in the amplitude and frequency of the systolic Ca^{2+} transient associated with physiological or pathological stimulation are brought about by catecholamines acting through β -adrenergic receptors. Catecholamines increase the magnitude and frequency of the Ca^{2+} transients via stimulatory G-protein to trigger the formation of cyclic adenosine monophosphate (cAMP), leading to activation of the protein kinase A (PKA)- dependent phosphorylation cascade and intensified calcium handling (cAMP-dependent positive inotropy). Once activated, PKA directly phosphorylates nodal Ca^{2+} regulatory proteins such as the L- type voltage-dependent calcium channels (VDCC) (increased Ca^{2+} influx), the ryanodine receptor (increased SR Ca^{2+} release), and phospholamban (PLN, increased SERCA2 activity) as a means of enhancing total intracellular Ca^{2+} release (MacLennan D. 2003, Colucci W. et al. 2000). Sympathetic stimulation increases the efficiency of EC gain, amplitude of calcium signal and consequently cardiac output, using the following potential point's controls in self-propagating calcium signal. First, the PKA-dependent phosphorylation increases magnitude of the calcium influx through L-type VDCC, i.e. the larger the influx the greater the "trigger", leading to increased RYR2 openings. Second, the Ca^{2+} content of the SR determines the amount of calcium released when a given number of RYRs are operative. B-Adrenergic stimulation results in PKA-mediated phosphorylation of the L-type voltage –dependent calcium channels and enhances the SR Ca^{2+} content. It is well established that acute signaling through both β1- and β2-adrenergic receptors result in PKA-mediated phosphorylation of nodal Ca^{2+} handling proteins, leading to augmentation in the Ca^{2+} transient and contractility (Petrashevskaya N. et al. 2008).

That Ca^{2+} serves as a primary effector of β-adrenergic receptor–induced heart failure was demonstrated by a rescue of disease associated with β1-adrenergic receptor overexpression through simultaneous deletion of the *PLN* gene (Engelhardt S. et al. 2004). This funding suggests that simple removal of diastolic Ca^{2+} caused by enhanced β-adrenergic receptor signaling was sufficient to rescue heart failure. These results also suggest that while β-adrenergic receptors may be desensitized in heart failure, they contain sufficient functional capacity to promote Ca^{2+} overload and enhance heart failure progression.

The calcium signal is a measure of stimulated electrical activity of the cell and can provide feedback control to the chemical messenger system (cAMP) that generates it. Evolution has put in place elaborate mechanisms for ensuring that these two major second messengers systems are tightly integrated (Cooper D. et al. 1995). The ability of numerous receptor types to couple to the inhibition and stimulation of AC activity was the first indication of the complex nature of the control of cAMP production. Existence of multiple types of AC s with different regulatory characteristics, make it a dynamic site of signal interaction from G-protein coupled receptors. There is considerable cross-talk between different second messenger system agents that mobilize intracellular calcium via activation of the PLC pathway, which can inhibit cAMP accumulation by directly inhibiting AC activity by protein kinase C (PKC). Many isoforms of AC, including types 5 and 6, are inhibited by Gi-linked receptors (Chakraborti S. et al. 2000). At the cellular level, cAMP isoforms are discretely localized to respond rapidly and selectively to Ca2+ entering the cell. In addition, types 5/6 AC which are predominant in the myocardium, can be effectively inhibited by small elevations in intracellular Ca^{2+} , providing a mechanism for cross talk between calcium-and cAMP-dependent signaling pathways (Muth J. et al. 2001, Yu H. et al. 1993), both of which are important regulators of cardiac contractility. Further, we should like to emphasize the importance of several calcium mobilizing and sensing mechanisms that provide a fine tuned cross-talk regulation of the cAMP and amplitude of calcium signaling system that are important in hypertrophied heart. The physiological interaction between cAMP producing and calcium mobilizing mechanisms (Na^{+}/Ca^{2+} exchanger) and the voltage dependent L-type VDCC is well established and may have an importance in a treatment of patients with heart failure. The relaxation of ventricular myocardium is reduced in heart failure and correlates, in some cases, with a prolongation of the Ca^{2+} transients suggesting a decreased capacity of the myocardium to restore a low basal Ca^{2+} level. One factor that correlates with prolongation of the Ca^{2+} transient is a deficient generation of cAMP, as has been observed in heart failure. Cardiac glycosides, including digoxin, are common cAMP-independent therapeutic agents used to increase the force of contraction. The Na^{+}/Ca^{2+} exchange –mediated rise in Ca^{2+} may relate to these abnormal cellular responses by inhibiting cardiac AC activity. Indeed, intravenous administration of digoxin impaired the peak positive left ventricular contractile response to dobutamine in rabbit hearts. This impairment was not caused by changes in the β-adrenergic receptor number or binding affinity, or by changes in the functional activity of Gs-protein. It was associated with impairment in the cAMP-producing activity of the catalytic subunit of AC. These defects may reflect an inhibition of AC activity by a rise in Ca^{2+} mobilized via the Na^{+}/Ca^{2+} exchange. This implies that the positive inotropic effect of cAMP-dependent agents, such as a dobutamine, may be attenuated in patients with heart failure who receive cardiac glycosides (Nagai K. et al. 1996).

Important negative feedback system has evolved within the self-propagating calcium signal itself to regulate the intensity of calcium induced calcium release and limit excessive calcium influx through the calcium/calmodulin sensing mechanisms. Enhanced calcium signaling can trigger increases in the cellular efflux, such that the SR Ca^{2+} content will decrease and in turn will decrease the systolic calcium transient. In non-excitable cells, this feed- back regulation between SR calcium load and calcium influx/efflux involves the process of capacitative calcium entry, in which the state of filling of the endoplasmic reticulum controls Ca^{2+} entry across the surface membrane via so-called store channels (Bers D. et al. 2000). Such a mechanisms is not effective in cardiac muscle, because it would be swamped by the large sarcolemmal calcium fluxes. Instead, a similar function is obtained by allowing Ca^{2+} release from the SR to regulate the L-type current and the Na^{+}/Ca^{2+} exchanger (Wier W. et al. 1999, Eisner D. et al. 2000).

3. ROLE OF INOSITOL 1, 4,5-TRIPHOSPHATE SENSITIVE CALCIUM IN ADRENERGIC TRANSDUCTION

The signaling role of inositol 1, 4, 5 triphosphate (InsP3) via intracellular calcium mobilization has been established in many cell types and associated with secretion, neurotransmission, fertilization, cell motility, and cell death. InsP3-sensitive calcium releasing channels (InsP3R) are activated by the second messenger IP3 which is generated via stimulation of phospholipase C (PLCβ) by either Gq-protein linked receptors or receptors linked to the src family of tyrosine kinases via PLCgamma (for review Petrashevskaya N. et al. 2002). InsP3 triggers a spatially restricted Ca^{2+} release from ryanodine and caffeine insensitive intracellular stores. It is unlikely that InsP3R, which is about 50 -fold less abundant than RYR in cardiomyocytes, can play an important role in regulating global calcium and the amplitude of calcium signal, even when up-regulated in cardiac hypertrophy and failure. More likely, IP3 serves a specialized function in cardiomyocytes for regulating organelle membrane permeability (including the nuclear membrane) and controlling local calcium signaling that may play a role in regulating cell growth pathways.

InsP3-sensitive calcium release plays an important role in regulation of β-adrenergic transduction through mechanism of heterologous desensitization. Several PKC isoforms, including PKCα, have been shown to be activated by Ca^{2+} from InsP3-sensitive pool. PKC a can phosphorylate and activate G protein-coupled receptor kinase-2 (GRK2) in vitro (Chuang T. et al. 1995, Krasel C. et al. 2001), and GRK2 is known to be a critical regulator of ventricular function. GRK2 is a member of the family of serine-threonine kinases known as GRKs (Pitcher J. et al. 1998). β-ARs are a target of phosphorylation by GRK2 leading to recruitment of β-arrestin and subsequent desensitization (Lohse M. et al. 1990) and signaling through myocardial β-ARs plays a critical role in normal and compromised heart function. Dysfunctional β-AR signaling in congestive HF includes receptor down-regulation and impaired signaling through the remaining receptors, possibly due to enhanced activity of

GRK2, which has been shown to be elevated in human HF. Desensitization of agonist-occupied receptors by the primarily cytosolic GRK2 requires a membrane-targeting event before receptor phosphorylation by a direct physical interaction between residues within the C terminus of GRK2 and the dissociated, membrane-anchored βγ subunits of G proteins (Koch W. et al. 2000). Inhibition of G_q-mediated activation of PKC and subsequent up-regulation of GRK2 activity may play an important role in the efficacy of angiotensin-converting enzyme inhibitors in patients with HF.

4. STRUCTURAL AND FUNCTIONAL REMODELING OF L-TYPE CALCIUM CHANNELS IN HEART FAILURE

Both the cellular and molecular basis of altered Ca^{2+} homeostasis leading to a decreased calcium transient and loss of contractility of the failing cardiac myocytes has been examined. However, the complexity of normal calcium regulation makes it difficult to select one critical point in the complex nature of dysfunctional Ca^{2+} regulation leading to abnormal systolic and diastolic calcium cycling. Although it has been reported that the density of RYR2 receptors decreased in failing myocardium, this down-regulation was not compensated by an up-regulation of the InsP3 sensitive calcium releasing channel (Go L. et al. 1995). The change in the abundance of one specific Ca^{2+} regulatory protein, itself, probably cannot cause a modification in Ca^{2+} metabolism.

The functional and structural Ca^{2+} channel remodeling also occurs in human heart failure and by it provoke pathophysiologic consequences on cardiac morphology, contraction, or rhythm. The finding that peak Ica is comparable in failing and non-failing myocytes when measured using standard voltage-clamp techniques may be hiding counterbalancing changes in channel density, gating, and single channel current (Mukherjee R. et al. 1998, Leclercq F. et al. 1998). First, the altered conformation of the action potential in failing myocytes can have a large effect on the influx of Ca^{2+} through L-type voltage dependent calcium channel even if all the properties of the channels are unchanged (Sah et al. 2002, Winslow R. et al. 1999). Second, alternative isoforms of the pore-forming Cav1.2 channel have been demonstrated in the failing heart and additionally, changes in auxiliary subunit composition may have important modulatory roles (Yang Y. et al. 2000). However, it is likely that altered Ca^{2+} homeostasis in hypertrophied and failing cardiomyocytes results from impairments in channel mobilization rather than changes in channel density and whole cell current. Third, in normal adult ventricular myocytes, most L-type Ca^{2+} channels are concentrated in the t-tubule network and contribute to the synchronized initiation of EC coupling throughout the myocytes. Failing cardiomyocytes can have dramatic cellular remodeling with loss of t-tubule membranes. Thus, it is possible that the localization of L-type calcium channels may be quite different in failing myocytes than normal myocytes. Schroder et al. (Schroder F. et al. 1998) provided evidences that the number of functional channels and the open probability of the channels were markedly increased in failing human ventricular myocytes without changes in single-channel current, just as if channels have been stimulated by PKA. Chen et al. (Chen X. et al. 2002) have found that maximally stimulated by isoproterenol Ica was significantly less in failing myocytes, given the well-documented blunted response of failing hearts to β-adrenergic stimulation. Although this may be in part a result of multiple alterations along the β -adrenergic signaling cascade in failing hearts, Chen X. Et al. suggested that additional contributions were due to a reduced density of L-type calcium channels in the failing myocytes. When the authors bypassed the β -adrenergic receptor through adenylylcyclase by directly testing nonhydrolyzable dibutyryl-cAMP, there was still a smaller maximally stimulated current level in failing myocytes. The effect of maximally activating channels via a distinct mechanism using the dihydropyridine BayK 8644 showed that the maximally stimulated Ica density was reduced compared with non-failing myocytes. Fortunately, these changes are at least partially reversible as demonstrated in patient with cardiac assist device (Kamp T. et al. 2002).

In mammalian hearts, L-VDCCs are composed of an ion conducting pore ($Ca_V1.2$ or A1C), and two auxiliary subunits, a $\alpha_2\delta$ and a β-subunits. Most investigators agree that β-subunit diversity is of physiological and pathophysiological importance (Hullin R. et al. 1992, Takahashi S. et al. 2003, Singer D. et al. 1991, Groner F. et al. 2004, Colecraft H. et al. 2002, Murakami M. et al. 2003). Molecular and functional analyses have shown that in heart failure enhanced L-VDCC activity is accompanied by altered expression pattern of auxiliary L-VDCC beta-subunit gene products, especially β2 subunit (Haase H. et al. 1996, Hullin R. et al. 1999) suggesting that an altered β-subunit expression pattern is of functional relevance. These data support the concept that in human heart failure, alterations of auxiliary subunit play an important role and at least some types of heart failure can be considered as L-type VDCC β subunit channelopathies. Using heterologous recombination, we have shown that distinct subunit compositions of L-VDCC induce single-channel characteristics similar to the biophysical phenotype of "hyperphosphorylated" L-VDCC (Hullin R. et al. 2003). Protein expression of $Ca_V1.2$, $\alpha_2\delta$, low molecular weight β_1, and β_3 was similar in non-failing and failing human myocardium, but there was a significant up-regulation of β_2. There was no difference in gene expression of the $Ca_V1.2$, and the $\alpha_2\delta$ at the protein level. At least two β_1-subunit isoforms ($\beta_{1a, c}$), four β_2-subunit isoforms (β_{2a-d}), and two β_3-subunit isoforms ($\beta_{3a, trunc}$) are expressed at relevant levels in human myocardium. B1a $_{and}$ β1c are sequence-identical except for replacement of exon 7a by exon 7b in β_{1c}, and $\beta2a_{-d}$ isoforms differ only with respect to the N-terminal region (D1 domain). Quantization by real-time PCR revealed an increased expression of β_{1c} and all β_2 isoforms in heart failure, in accordance with the protein data.

Human tissue offers a limited choice of truly independent variables, such as time, disease stage and treatment options and therefore delineation of pathophysiological mechanisms in human heart is difficult because of wide inter-individual variance, including age, medication, state of disease. Animal models offer control of any relevant factor to test pathophysiological concepts. We analyzed β-subunit gene expression in both human non-failing and failing hearts as well as in transgenic mice overexpressing the human $Ca_V1.2$ (A1C) subunit (TG $Ca_V1.2$). The latter was chosen because of phenotypical characteristics common with human heart failure, *e.g.* early blunting of β-adrenergic signaling, slow progression towards hypertrophy and calcium overload in failing myocytes (Muth J. et al.1999, Muth J. et al. 2001).

At four months, TG mice showed a mild reduction in percentage fractional shortening (%FS) and heart rate, but other parameters did not change significantly. In contrast, at eight months and older, severe cardiac enlargement and marked cardiac dysfunction were evident. This late phenotype was associated with a progressive increase in the left ventricular mass, decrease in %FS, and increase in left ventricular end-diastolic dimension and left ventricular end-systolic dimension

Table. Echocardiographic and hemodynamic parameters in transgenic mices overexpressing the á1-subunit if L type voltage dependent calcium channels

	NTG	TG-á1OE	NTG	TG-á1OE
FS,%	50.4±1.9	43.6±2.6	47.7±2.3	31.7±2.8^
ESD, mm	1.4±0.09	1.7±0.11	1.6±0.08	280.23^
EDD, mm	2.8±0.12	3.0±0.08	2. 9±0.05	4.00±17^
AWT,mm	0.62 ±0.02	0.65±0.03	0.62±0.03	0.62±0.03^
PWT,mm	0.61±0.01	0.63±0.02	0.65±0.03	0.64±0.02^
h/r(PW/EDD-2)	0.4±0.014	0.43±0.02	0. 43±0.02	0. 32±0.02^
LVM, mg	47.7±4.5	54.3±3.9	52.8±4.4	87.2±6.1^
HR, beats/min	677±28	591±28.3*	690±14	522±12^
E time, ms	40.3±0.81	44.1±0.96*	38.4±1.1	49.1±1.6^
Vcf,circ/s	12.6±0.62	9.9±0.65*	12.4±0.66	67±0.65^
N(mice studied	7	7	9	17
Contraction properties				
+dP/dt,mmHg/s	3.836.7±147	5.012±225*	3.418±149	3.903±182*
TPP,ms/mmHg	0.46±0.03	0.37±0.02	0.42±0.01	0.41±0.02
Relaxation parameters				
-dP/dt, mmHg/s	2.922±158	2. 818±108	2.744±179	2.080±141*
Rt1/2, ms/mmHg	0.74±0.05	0.75±0.07	0.66±0.04	0.91±0.09*
N(mice studied)	6	6	5	4

AWT = anterior wall thickness; +dP/dt = maximal rate of left ventricular pressure development; −dP/dt = maximal rate of left ventricular decline; EDD = end-diastolic dimension; ESD = end-systolic dimension; E TIME = ejection time; FS = fractional shortening; HR = heart rate; LVM = left ventricular mass; Ntg = no transgenic; PWT = posterior wall thickness; R/T1/2 = time to half-relaxation; Tg = transgenic; TPP = time to peak pressure; Vice = mean velocity of circumferential fiber shortening.

Most importantly, in young (non-failing; "Adaptive State") TG $Ca_V1.2$ mice we previously found concordance of lowered β_2-subunit expression and decreased activity of single L-VDCC (Groner F. et al. 2004). An increase of single L-VDCC activity accompanied by enhanced expression of β_2-subunits was found when these mice have entered the failing state ("Maladaptive State" ≥ 9 months of age). In order to further elucidate the putative role of β_2-subunits in heart failure, human pathophysiology with respect to Ca^{2+}-channel function, was simulated using a genetic mouse approach. The single channel phenotype of human heart failure can be reproduced in the mouse heart by short-term overexpression of a β_{2a}-subunit (ti_n β_{2a}). As a novel and first approach to induce an increased β_2-subunit overexpression in intact animals, rather than in isolated cells (Colecraft H. et al. 2002) mouse model of cardio specific inducible β2a-subunit expression (ti_n β_{2a}) was generated. Cardiac overexpression of β_{2a}-

subunits was induced *in vivo* for a period of 4 weeks in mice through continuous application of the selective inducing drug tebufenozide via osmotic pumps (Hoppe U. et al. 2000, Beetz N. et al. 2009). It is important to emphasize that the continued induction of only the β_2-subunit transgene was totally sufficient to resemble the Ca^{2+}-channel phenotype of human heart failure: little change (even unchanged at positive test potentials) of whole-cell current density, leftward shift of voltage-dependent activation, marked enhancement of single-channel activity, and blunting of the Ca^{2+}-current response towards phosphorylating conditions (Schroder F. et al. 1998).

Induction of β_2-overexpression in this mouse model did not affect overall single L-VDCC gating significantly. A 1:1-stoichiometry of pore-forming α_1- and auxiliary β-subunit may be sufficient for modulation of channel gating, and therefore most calcium channel pores are saturated with native β-subunits in the induced ti_n β_{2a}. Mean closed time was lower in induced ti_n β_{2a} suggesting that a portion of overexpressed β_{2a} exerts functional action similar to the recombinant channel.

To prove the concept that β_2-subunit expression underlies the activity of single L-VDCC of the heart-failure phenotype, ti_n β_{2a} with transgenic $Ca_V1.2$ mice were crossbred and investigated. Induction of β_{2a}-subunit gene expression in the young double transgenic mice (TG $Ca_V1.2 \times tg_{ind}$ β_{2a}) led to a premature increase of single L-VDCC activity (Hullin R. et al. 2007). These data are consistent with idea that Ca^{2+}-current in human heart failure is carried by a reduced number of channels, each individually contributing more current at the expense of a reduced regulatory bandwidth. These findings do not resolve the problem why in failing heart single channels, at least those located at the surface sarcolemma, are twice as active as would be predicted from largely unchanged whole-cell currents, with no direct evidence of a reduced expression of channel pore protein (Hullin R. et al. 2007). Nor do they completely elucidate the mechanistic basis of hyperactivity of single channels (hyper-phosphorylation, change in subunit isoform composition, or a combination of both).

In vivo overexpression of β_2-subunits did not affect systolic contractile function and did not cause significant changes in transgenic $Ca_V1.2$ or double-transgenic hearts. This argues against a major adaptive role of β_2-subunit overexpression at least in young animals (4–5 months) as studied here, i.e. during a stage where heart failure is not yet developed. Lower blood pressure in transgenic $Ca_V1.2$ and double-transgenic mice may be explained by the significant bradycardia, as left ventricular maximal rate of pressure rise dp/dt_{max} was not altered in these mice. In this context, it is intriguing to speculate that reduced ventricular relaxation in $Ca_V1.2$ transgenic mice may be due to diastolic dysfunction associated with ventricular interstitial fibrosis.

These data prove that the entire spectrum of functional Ca^{2+}-channel remodeling as it occurs in human heart failure can be reproduced by continuing overexpression of a β_2-subunit alone. This serves as the basis to address the important question whether β_2-subunit overexpression is an adaptive or maladaptive phenomenon, and whether interference with structural remodeling would constitute not only a technically feasible but also a therapeutically promising approach.

5. Ca^{2+}-Dependent Pro-Hypertrophic Signaling Pathways

Increases in intracellular calcium induces pathological cardiac hypertrophy, however, the source of the hypertrophic Ca^{2+} is not clearly identified. Persistent increase in calcium influx through L- type voltage- dependent calcium channels, the primary Ca^{2+} influx pathway in the heart, is sufficient to initiate pro-hypertrophic signaling cascades. Conversely, and of more pathophysiological relevance, *in vivo* knockdown of β_2-subunits in a rat model of aortic banding reduced Ca^{2+}-current density and attenuated cardiac hypertrophy (Gingolani I. et al. 2007). L-type type calcium channel blockers eliminate neonatal rats ventricular myocytes hypertrophy induced by many neurohormones and stretch (Ikeda K. et al. 2000, Zho Y. et al. 2002). T-type calcium channels and transient receptor potential channels have all been proposed to contribute to the pool of Ca2+ that activated hypertrophic pathways (Onohara N. et al. 2006, Chiang C. et al. 2009, Heineke J. et al. 2006)

Acute β2-subunit overexpression has been shown to cause cell death and apoptosis in isolated myocytes and life-long overexpression in a transgenic mouse model resulted in hypertrophy, failure, and premature death, albeit under conditions where whole-cell current density was vastly elevated (Hiroyuki Nakayama et al. 2007). This finding confirms that chronic increases in calcium load through elevated L-type channel expression and function in the sarcolemma can lead to hypertrophy and failure. Transgenic mouse models with low and high expression level of the β2a subunit of Ca^{2+} channels had increased heart weight/body weight ratio, posterior wall and intraventricular septal thickness, tissue fibrosis. Using this transgenic model, it has been demonstrated that Ca^{2+} influx–induced myocyte necrosis might be a significant contributor to heart failure. Necrosis occurs in most tissues following ischemic injury or Ca2+ overload in combination with depletion of high-energy phosphates such as ATP, resulting in mitochondrial swelling and rupture (Danial M. et al.2004, Zamzami N. et al.2001, Cropmton M. et al. 2002). Necrosis associated with mitochondrial swelling and rupture is not thought to be a strictly passive process, but instead can be regulated at the level of the MPT pore through the activity of cyclophilin D (Halestrap A. et al. 2006). Heart failure is a complex and progressive syndrome that often culminates in death by arrhythmias or hemodynamic instability. One significant mechanism underlying its progressive nature is a stochastic loss of cardiac myocytes that eventually renders the heart with too few cells to properly function (Kajstura J. et al. 1996). Indeed, alterations in Ca^{2+}, either global or within select microenvironments, can activate Ca^{2+}/calmodulin-dependent kinase in the heart and lead to further alterations in Ca^{2+} handling that can induce myocyte apoptosis and heart failure (Zhu W. et al.2003). This cumulative loss of myocytes in heart failure has been generally attributed to an apoptotic process (Foo R.S. et al. 2005), suggesting the use of pharmacologic agents that antagonize apoptotic effector proteins such as caspases, pro-death Bcl-2 family members, or nucleases (Wencker D. et al. 2003).

In response to disease causing stimuli, the myocardium is influenced by neuroendocrine secreted growth factors and/or cytokines that induce ventricular remodeling, hypertrophic enlargement of myocytes, and alteration in the viability of myocytes. Many of this neuroendocrine factors such as angiotensin II and endothelin-1 alters calcium homeostasis through G-protein coupled receptors on cardiac myocytes to induce phospholipase C (PLC) activation, which in turn leads to generate InsP3 and diacylglycerol (DAG). In most cell types, InsP3 generation in turn leads to release of Ca^{2+} from the endoplasmic reticulum through the InsP3R channel (Woodcock E. et al. 2005).

How is a hypertrophic Ca2+ signal can be discriminated from the repetitive Ca^{2+} rises underlying contraction? How is the hypertrophic Ca2+ dependent signal at the surface membrane transduced to the nucleus? Two scenarios exists to understand how changes in cardiomyocyte Ca2+ concentration could regulate specific signaling pathway in the backdrop of cyclic calcium transient. The first scenario postulates the existence of specialized cellular microdomains in which calcium concentration is locally regulated and sensed by macro-molecular signaling complexes. According to this scenario, systolic Ca^{2+} transient has no effect on hypertrophic signaling pathways. The second possibility is that the same pool of contractile calcium that controls cardiac contractility underlies activation of several specific signaling pathways (Berns DM. et al. 2005).

Ca^{2+} regulates many hypertrophic pathways including well-known calcineurin/NFAT, CaMKII/HDAC and PKCά . Hypertrophic pathways including CaN/NFAT, CaMKII/HDAC, PKC and MAPK pathways are all regulated by Ca^{2+} (Heineke J. et al. 2000) but calcium activation requirements are unique for each pathway. It is believed that differences in amplitude, frequency, duration, or subcellular localization of the Ca^{2+} signals elicited by the hypertrophic agents may represent the biological code for pro-hypertrophic action (Colella M. et al. 2008). Understanding the pathophysiologic relevance and relative contribution of specific Ca^{2+} -dependent pro-hypertrophic pathway is necessary to developing new molecular targets for complex intervention in the treatment of heart hypertrophy and failure.

A. Calmodulin- calmodulin-dependent kinase

Calmodulin/Calmodulin-dependent kinase (CaMK) is ubiquitous mediators of Ca^{2+} signaling. This multifunctional serine/threonine kinase family, consisting of CaMKI, -II and – IV, has wide tissue distribution, can phosphorylate multiple substrates and regulate numerous cellular function. CaMK is activated by high amplitude Ca^{2+} spikes and its activity is also dependent on Ca^{2+} spike frequency (Bodi I. et al. 2005). There are numerous pathways that can regulate calcium homeostasis and transcriptional activity through calmodulin/CaMK (Ho N. et al. 1996, Wegner M. et al. 1992).

CaMK is known regulator of calcium-dependent inactivation of L-type calcium channel. The cardiac L-type-voltage dependent calcium channel display long-lasting openings and minor voltage-dependent inactivation component. In the heart, Ca^{2+}- dependent inactivation is compatible with the length of the Ca^{2+}-mediated plateau phase in the action potential. Therefore, Ca^{2+} -induced inhibition of the cardiac L-type voltage –dependent calcium channels plays a critical role in controlling calcium entry and downstream signal transduction as well as ensuring that contraction and relaxation cycles of the heart muscle are coordinated. (Peterson B. et al. 1999). Several lines of evidence suggest that the Ca^{2+}-binding protein calmodulin (CaM) is a critical sensor in mediating the inactivation process of L-type calcium channels. In the C terminal tail of the pore forming άlc subunit of L-type VDCC, there is a Ca^{2+}-dependent CaM-binding isoleucine-glutamine (IQ) motif that has been implicated in auto-regulation (Kim J. et al. 2004). Ca^{2+}-CaM-CaMKII-L-type-VDCC crosstalk has a central role in contractility and Ca^{2+} homeostasis and ensures tightly-controlled calcium influx.

Substantial evidence support involvement of calmodulin/CaMK in cardiac hypertrophy and failure through regulation of gene transcription. The data presented by Wu et al (Wu X. et al. 2006) reveled a novel CaMKII-dependent local control system for excitation-transcription coupling in adult ventricular myocytes and established a paradigm whereby a specialized pool of Ca^{2+} can regulate a hypertrophic circuit independent of contractile calcium without being distracted by beat-to-beat calcium transient. This pathway is believed to be important in cardiac transcriptional regulation and demonstrate an extremely important way by which myocytes can use CaMKII activated by pro-hypertrophic neurohumoral stimuli. The physiological hypertrophic agonist endothelin -1 (ET-1), angiotensin, phenylephryne coupled to Gq-proteins activate phospholipases and cause production of both inositol 1,4,5 triophosphate (InsP3) and diacylglycerol (DAG). DAG activates several PKC isoforms that causes many downstream functional and transcriptional effects (Dorn G. et al. 2005). Inositol 1,4,5-trisphosphate receptors (InsP3R) are ubiquitous intracellular Ca^{2+} release channels. They are present in cardiac myocytes at lower levels than the related ryanodine receptor (RyR), which is the main source of Ca^{2+} in excitation- contraction coupling. The type 2 InsP3R is the predominant subtype in cardiac myocytes located mainly in the nuclear envelope where they complex with CaMKII. In cardiac myocytes InsP3R2 is physically associated with CaMKII and serves a more specialized role in providing a local pool of Ca^{2+} to activate this kinase. Ca^{2+} from an InsP3 R-dependent pool produced by ET-1 induces activation of phospholipases and causes very local Ca^{2+} release via InsP3R2 in the nuclear envelope, which in turns activates CaMKII, CaMK phosphorylates type II histone deacetylases (HDAC 4,5,7,9) and causes consequent HDAC5 nuclear export, and activation of MEF-2 dependent transcription. These HDAC s normally repress transcriptional activation driven by myocytes enhancer factor -2 (MEF2) and favor condensed DNA. When HDAC is phosphorylated in response to neurohumoral stimuli, it is exported from the nucleus in association with the chaperone protein 14-3-3, MEF-2 is de-repressed, and a hypertrophic program of cardiac gene expression is activated (Olson E. et al. 2003, Zhang C. et al. 2002). This movement of HDAC5 in association with InsP3R-dependent Ca^{2+} mobilization was partially mediated by CaMK activation since pharmacological inhibition blocked approximately half of the HDAC5 nuclear export.

Moreover, CaMKII is directly complexed with InsP3R in cardiac myocytes, suggesting a mechanisms whereby CaMKII might only respond to exceedingly high Ca^{2+} levels in the microenvironment of the InsP3R and insensitive to total intracellular Ca^{2+} levels. More convincingly, HDAC5 nuclear export after endothelin-1 stimulation is completely absent in adult cardiac myocytes from InsP3 gene –targeted mice. Activation of InsP3R-associated CaMKII can phosphorylate the InsP3R2 and decrease its open probability (Bare D. et al. 2005). This may be negative feedback mechanisms, such that Ca^{2+} released from the InsP3R activates local CaMKII, which feeds back to shut off Ca^{2+} release from the nuclear envelope. The activated CaMKII can then phosphorylate other targets. The Ca^{2+} involved in this activation is relatively insulated from the Ca^{2+} transients associated with every heart beat and responses to neurohumoral stimuli and cytosolic calcium transients initiated through L-type voltage –dependent calcium channels are distinct. This pathway may also compliment other Ca^{2+} and CaM-dependent transcriptional factors, such as the calcineurin/NFAT pathway, which responds to very different local or global Ca^{2+} signal than described for CaMKII/HDAC5.

CaMKII may activate transcription via phosphorylating cAMP response element-binding protein (CREB), which is a major downstream target for CaMKII. CREB normally regulates the transcription of target genes that encode contractile proteins, proteins involved in generating energy, and proteins required for cardiac myocytes growth and viability. A variety of intracellular signaling pathways may be involved in CREB phosphorylation and activation, including PKA and CaMKII in response to elevation in Ca^{2+}. Therapies increasing CREB activity in the failing heart have been proposed to slow the progression of heart failure (Muller F. et al. 2000).

B. Protein kinase C

The protein-kinase C (PKC) family of Ca^{2+} and/or lipid-activated serine/threonine kinases function downstream of many signal transduction pathways (Liu Q. et al. 2009, Liu Q. et al. 2010). The PKC family comprises 12 different isozymes, which are broadly classified by their activation characteristics. The conventional PKC isozymes (α, βI, βII ,γ) are Ca^{2+} activated and lipid-activated, whereas the novel isozymes (ε, θ, ή, δ) and atypical isozymes (ζ, τ, μ, ν) are Ca^{2+} independent but activated by distinct lipids. PKCά is the predominant PKC isozyme functioning as a proximal regulator of Ca^{2+} handling. PKCά activation and increase in PKCα expression is associated with hypertrophy and dilated cardiomyopathy, ischemic injury, and mitogen stimulation. PKCά is normally activated by increases in intracellular calcium concentration and/or lipid signaling mediators. For the classical PKC isozymes, binding of Ca^{2+} and phosphotilserine to the C2 domain leads to increased membrane association. For all PKC isoforms, membrane translocation provides mechanisms to regulate substrate access through docking complexes.

A number of PKC ά -dependent molecular mechanisms have been investigated to offer protection from heart failure by PKC inhibition. One of likely mechanisms associated with protection from heart failure is negative modulation of cardiac contractility mediated by PKC ά. Strong evidence exist that PKC ά is a potential regulators of Ca^{2+} handling and cardiomyocyte contractility. This PKC ά dependent regulatory pathway affords mechanisms for feed-back inhibition. Sustained augmentation of cardiac inotropy associated with higher calcium transients or with neuro-endocrine stress signals activates PKC ά in the heart, resulting in a negative inotropic effect. PKC ά inhibition enhances cardiac contractility through sarcoplasmic reticulum Ca^{2+} loading. Using this mechanisms, PKC ά regulates transduction pathways initiated by β-adrenergic receptors and cAMP through direct phosphorylation of inhibitor 1 (I-I) at serine67, resulting in altered phosphatase-1 (PP1) activity, leading to greater phospholamban (PLN) dephosphorylation and less activity of the sarcoplasmic reticulum Ca^{2+} ATPase (SERCA2). Alteration in PLN phosphorylation in turn regulates Ca^{2+} loading and the magnitude of Ca^{2+} transients. Decreased SERCA activity results in reduced SR calcium loading and reduced contractility. There is a unique aspect associated with PKC inhibition, whether resulting from gene ablation or the pharmacologic compounds. PKC ά inhibition only moderately increases contractility and calcium transients. Beta-adrenergic agonist have been known to have a stronger effect on cardiac inotropy, that have been associated with increased risks of more rapid decompensation to heart failure patients (Bristow M. 2000). Pharmacological inhibition of PKC ά activity functions at the level of SR Ca^{2+} handling, possibly providing a safer inotropic effect and presenting a more refined target in which cardiac contractility is augmented within physiologic limits.

Inhibition of PKC ά may also benefit a failing myocardium independent of contractility, since PKC ά is a mediator of reactive signaling that participates in hypertrophy and decompensation. Emphasis has been placed on the ά -isoform of PKC because it is elevated in human heart failure (Bowling N. et al. 1999). We found (Muth J. et al. 2001) a significant activation of PKC ά in transgenic mice with persistent increase in calcium influx due to overexpression of pore-forming subunit of L-type VDCC, particularly at 2 months, which is before the development of frank hypertrophy. This funding suggests an important " trigger" role for PKC in the induction of the hypertrophic program initiated by increased calcium influx through the L-type VDCC. Thus, it is likely that PKC ά , which is Ca^{2+}-dependent, is activated by the sustained increase of the Ca^{2+} current early on in our transgenic mice and serves as the initiator of the hypertrophic gene program. Additional PKC substrates important in cardiac disease include the β -adrenergic receptor (β AR) and a key β -adrenoreceptor modulator β-AR kinase (BARK or GRK2). Phosphorylation of these substrates via PKC decreases β AR signaling. Because 2-month-old transgenic animals with pore forming subunit of L-type VDCC showed a striking loss of the inotropic response to isoproterenol (Muth J. et al.1999), it is likely that PKC also underlies the blunting of the β -adrenergic signaling pathway observed in this transgenic model. Furthermore, β -adrenergic desensitization and elevation of β ARK1 levels precede the development of heart failure in both transgenic mouse models (Cho M. et al. 1999) and infarcted rabbits (White D. et al. 2000). Thus the web of signaling pathways including Ca2+ , PKC, BARK, and the β-AR are intimate players in the disease development of this model (Figure 1).

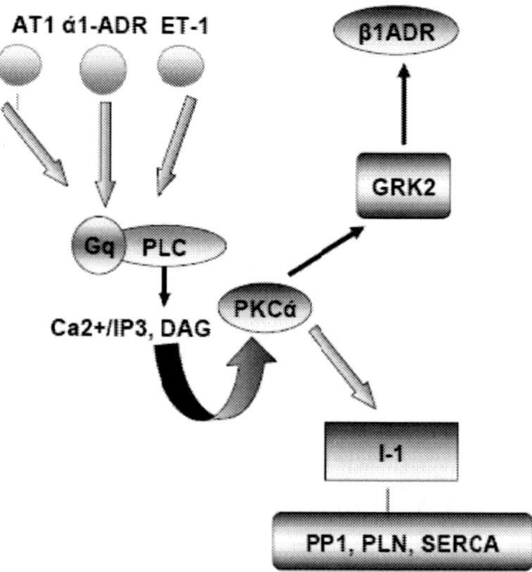

Figure 1. Diagram of signaling in a cardiac myocyte showing how PKCα becomes activated by Gαq-coupled receptors leading to phospholipase C (PLC) activation and the liberation of DAG and Ca2+ . Once activated, PKCα has several mechanisms whereby it can alter cardiac function (arrows). The activated β1-AR is desensitized when they are phosphorylated by β-AR kinase 1 (BARK1, GRK2).

PKCα also appears to directly phosphorylate key myofilament proteins including cardiac troponin I (cTnI), cTnT, titin, and myosin binding protein C, which leads to decreased myofilament Ca2+ sensitivity and reduced contractility in myocytes (Belin R. et al. 2007). Moreover, PKCα has also been shown to phosphorylate the pore-forming α1c subunit of the L-type Ca2+ channel, an effect that could alter contractility as well (Yang L. et al. 2009). Similarly, PKCβI, βII, and γ can also phosphorylate serine residues in the A1C subunit. These data further suggest the possible promiscuous nature of the cPKCs in the regulation of protein phosphorylation, Ca^{2+} cycling, and cell contractility. Despite these similarities, PKCα functions are unique from PKCβ and γ in altering contractility because adult cardiac myocytes from PKCβ overexpressing transgenic mice showed increased Ca^{2+} transients and increased contractility (Huang L et al. 2001), while PKCα overexpressing transgenic mice showed depressed cardiac contractility. Moreover, myocytes from adult PKCα null hearts showed increased contractility and augmented Ca^{2+} transients.

Most clinical trials of pharmacological positive inotropes in human heart failure have shown adverse outcomes and increase in mortality. On the opposite site, enhancement in cardiac contractility associated with PKC ά gene deletion protected against pressure-overload induced heart failure and dilated cardiomyopathy (Braz J. et al. 2004). Acute inhibition of the conventional PKCά isoform with bisindolylmaleimide compounds, Ro-32-0432 and Ro-31-8220 significantly augmented cardiac contractility and also acutely increased cardiac contractility in different models of heart failure. Adenoviral-mediated gene therapy with a dominant-negative PKCά cDNA rescued heart failure in a chronic rat model of post-infarction cardiomyopathy (Hambleton M. et al. 2006). Another PKCα/β inhibitor, ruboxistaurin, has been through late stage clinical trials for diabetic macular edema and shown to be well tolerated and hence was extensively analyzed in both mouse and rat models of heart failure (The PKC-DRS Study group. 2007). Although ruboxistaurin was originally reported to be PKCβ selective, it has been determined to be equally selective for PKCα (IC50 of 14 nmol/L for PKCα versus 19 nmol/L for PKCβII). Moreover, given that PKCα protein levels are much higher than PKCβ in the human and mouse hearts (Hambleton M. et al. 2006), it further suggests that ruboxistaurin functions predominantly through a PKCα-dependent mechanism. Ruboxistaurin increased baseline contractility by 28% in rats with acute infusion. Acute infusion of ruboxistaurin also augmented cardiac contractility in wild type and PKCβγ–/– mice but not PKCα–/– mice. These results indicate that ruboxistaurin enhances cardiac function specifically through effects on PKCα but not PKCβ or PKCγ. In other words, all of the protective effects observed with ruboxistaurin in rodent models of heart disease are predominately dependent on PKCα inhibition. Ruboxistaurin prevented death in wild type mice throughout 10 weeks of pressure overload stimulation, reduced ventricular dilation, enhanced ventricular performance, reduced fibrosis, and reduced pulmonary edema comparable to or better tan β AR -blocker metaprolol treatment.

Regardless of the mechanisms, PKCά is an attractive pharmacologic target for treating human heart failure. PKCα inhibition has a prominent effect on SR Ca^{2+} cycling and the myofilament proteins as a means for altering cardiac contractility. These mechanisms of action are significantly downstream of how traditional β-adrenergic receptor agonists function and hence might bypass the negative effects of traditional inotropes that promote arrhythmia and myocyte death.

C. Ca2+ activated protein phosphatase calcineurin (PP2B)

Calcineurin is a Ca^{2+}/calmodulin activated serine/threonine phosphatase, which plays a significant role in cardiac hypertrophy as a sensing molecule that links alterations in calcium handling and the genetic program of hypertrophic growth. Calcineurin, as other calmodulin-dependent enzymes, is activated by prolonged, low amplitude increases in basal concentration of calcium and signals hypertrophy by dephosphorylation of transcription factors of the NFAT (nuclear factor of activated T cells) family, which then translocates to the nucleus and interacts with cardiac restricted zinc finger transcriptional factor GATA4 to activated genetic reprogramming and initiation of the hypertrophic transcriptional response. Expression of the catalytic subunit of calcineurin in transgenic mice was shown to induce profound hypertrophy with a two- to four-fold increase in the heart size, which rapidly progresses to dilated heart failure (Molkentin J. et al. 1998). Calcineurin triggers a signal that is sufficient to activate a physiological adaptive response that may compensate and protect the heart against functional insufficiency under conditions of elevated after-load or disturbed contractility secondary to abnormal myofibrillar/cytoskeletal architecture. Reduced spontaneous frequency of contraction in calcineurin- overexpressing mouse heart can be considered as one possible adaptation leading to a decreased energy demand by increasing the period of diastole. On the other hand, the reduced heart rate that we observed may be a direct consequence of a shifted balance of phosphorylation/dephosphorylation systems providing continuous modulation of excitability and firing properties (Petrashevskaya N. et al. 2002, Pedarzani P. et al. 1998, Frace A.M. et al. 1993). Reactive functional compensation and structural remodeling in the calcineurin-overexpressed hearts appears to reflect an important compensatory mechanism such that, without this adaptation, the heart could not sustain the pressure overload and might decompensate to failure more rapidly. As reported by Meguro et al. (Meguro T. et al. 1999) the incidence of decompensation leading to heart failure increased with the inhibitory action of cyclosporine on the development of LVH (left ventricular hypertrophy). As recently shown, concentric hypertrophy and decompensated heart failure share a common signaling pathway, but calcineurin activation is predominant in stable hypertrophy versus heart failure (Haq S. et al. 2001).

If the sufficiency of calcineurin to promote cardiac hypertrophy have been demonstrated in vivo and in vitro and have been long known (Molkentin J. et al. 1998), the source of intracellular calcium for reactive calcineurin-dependent signaling is obscure.

Calcineurin/NFAT axis has been proposed to act as integrator of the Ca^{2+} signals. A number of investigators suggested that contractile calcium entry through L-type calcium channel directly controls calcineurin-NFAT signaling in cardiac myocytes. In neonatal rat ventricular cardiac myocytes, angiotensin produced an increase in the frequency of contractile calcium transients, leading to an increase in the time-averaged bulk cytosolic calcium. Similarly, addition of KCl to the culture media to depolarize the plasma membrane increased Ca^{2+} transient frequency and produced a hypertrophic response measured by cell area (Haq S. et al. 2001). Hypertrophy associated with KCL was blocked by an inhibitor of the L-type Ca^{2+} channel, which reduces calcium influx, but not by a generalized inhibitor of contraction. Even increasing Ca^{2+} transient frequency by field stimulation was sufficient to activate NFAT nuclear translocation. However, a direct link between increased contractile calcium and pathophysiologic hypertrophy still cannot be made in vivo.

Persistent changes in the amplitude and duration of the systolic Ca^{2+} transient that are specific to pathological stress could induce pathologic hypertrophy signaling. The mechanisms of NFAT activation are not clear, because of the very brief duration of each single calcium spike in cardiac cells. One regulatory mechanism for compartmentalizing Ca^{2+} outside of excitation-contraction complex has been investigated. The voltage-regulated L – type calcium channel, not associated with junctional complex and Ca^{2+}-induced calcium release, but dedicated to special membrane domains lipid rafts have been suggested to mediate a signaling process. Calcineurin activation and NFAT translocation is more dependent on the change in cytosolic Ca^{2+} while HDAC translocation is more related to the increase of sarcoplasmic reticulum - nuclear envelope Ca2+. It is likely that neuroendocrine or stretch-dependent signaling receptors directly communicate with this presumed Ca^{2+} microdomain to induce the hypertrophic response and activation of Ca^{2+}-dependent signaling effects. Elucidation of additional Ca^{2+} signaling microdomains in adult cardiac myocytes is important in resolving how the myocytes dissect signaling versus contractile calcium.

In human heart failure, there are complex changes in the regulation of contractile calcium, with decreased rates of calcium reuptake by the SR, prolongation of calcium transient, increased diastolic Ca^{2+}. Calcineurin-NFAT can be activated in this scenario (Haq S. et al. 2001).

An alternative possibility identifies the NFAT shuttle as a typical example of "integrative tracking". According to this theory, calcineurin/NFAT axis has been proposed to act as integrator of the Ca^{2+} signals. During a single Ca^{2+} oscillation of 200-ms, only very small fraction of NFAT may be dephosphorylated by calcineurin. If the interval between two successive Ca^{2+} spikes is shortened, the balance between phosphorylation /dephosphorylation will be shifted toward the dephosphorylation, and dephosphorylated NFAT will accumulate in the cell. This kinetic scheme may explain not only frequency dependent NFAT nuclear translocation, but may also be applied to stimuli that may affect the amplitude rather than frequency of calcium transient. In this case, a larger Ca^{2+} increase will activate more calcineurin molecules and the shift toward more dephosphorylated NFAT will accumulate if the episodes occur frequently in the same cells. This model named "integrative tracking" proposes that each calcium transient has a small effect on itself, but individual transients can be integrated over time to provide a significant change in cellular process (Colella M. et al. 2008).

Calcineurin may also be a player contributing to blunted β-adrenergic transduction in cardiac hypertrophy. Suppression of cyclic adenosine monophosphate (cAMP) signaling is required for T-cell activation by calcineurin. Several transcription factors, including inducible cAMP early repressor, NF-KB, c-Jun N-terminal kinase and NFAT, have been identified as potential mediators of negative regulations of cAMP (Bodor J. et al. 1998, Neumann M. et al. 1995). The mechanism of inhibition of cAMP signaling in T cells appears to be mediated by a transcriptional induction of phosphodiesterase 7 (Li L. et al.1999). In addition, in excitable cells, initiation of the cAMP signal by adenylyl cyclase (AC) may be dynamically controlled by Ca^{2+}/calcineurin and thus provide evidence for a novel mode of tuning the cAMP-dependent cellular response by protein phosphorylation/dephosphorylation cascades (Antoni F. et al. 1998). The effect of calcineurin on cAMP synthesis appears to be associated with the expression of a novel AC isotype IX. Human AC IX is expressed in vital organs, including the heart. In rodents AC IX is expressed in a wide variety of tissues including brain, heart, skeletal muscle and pancreas but its role in physiological control remains to be explored (Paterson J. et al. 1995, Premont R. et al. 1996).

Using AC as a dynamic target of signal interaction, calcineurin has been shown to act as a $Ca^{2+/}$calmodulin operated feedback inhibitor of the β-adrenergic receptor-evoked cAMP response in different cell types (Paterson J.et al. 1995, Premont R. et al. 1996) .Very little is known whether the latter represents a physiologically relevant mechanism that might contribute to the well-known blunted β-adrenergic responsiveness in animal models of cardiac hypertrophy and failure as well as in human pathological conditions (Koch W. et al. 2000, Gao M. et al. 1998, Gao M. et al.1999). The functional versatility of structurally different ACs allows for simultaneous activation/or inhibition by multiple mechanisms. Calcium itself can provide a negative regulation of cAMP-dependent contractile responses by interacting with $Mg^{2+/}Ca^{2+}$ sites on the enzyme.

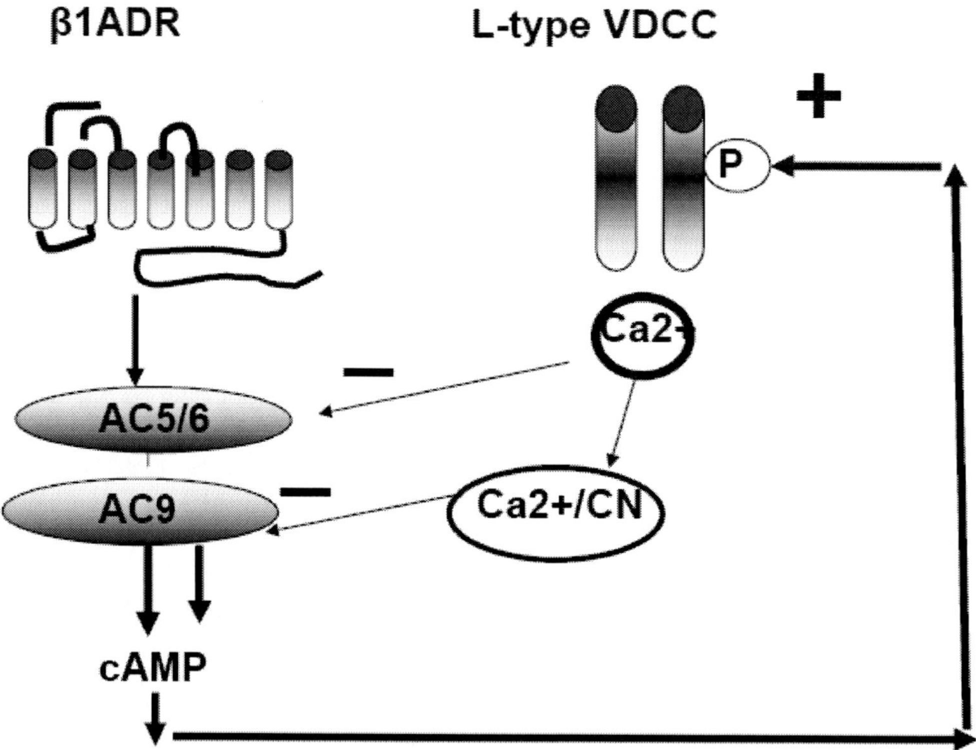

Figure 2. Intracellular pathway for Ca2+ dependent negative feed- back mechanism over sympathetic stimulation through AC5/6 and AC9. β1-adrenergic stimulation initiates positive inotropic effect via AC5/6. Activation of AC5/6 produces an increase in the cAMP, which phosphorylates the α1-subunit of the L-type calcium channels. AC5/6 ia not affected by Ca2+/calmodulin, but can be inhibited by intracellular calcium. Inhibition of AC9, through calmodulin/calcineurin (Ca2+/calcineurin) is hypothetical Ca2+-dependent pathway to inhibit cAMP production.

In excitable cells, calcineurin plays a role in the negative feedback regulation of Ca^{2+} entry through the LVDCC weakening intracellular Ca^{2+} signaling (Wang J. ET al.1997). One of the possible pathways to increase intracellular pool of calcium is AC/cAMP/PKA-dependent phosphorylation of calcium-handling proteins. On the other hand, phosphorylation of PKA responsive sites on the NH2-terminal region of NFAT may cause export of NFAT from the nuclei to the cytoplasm and therefore causes inhibition of NFAT-mediated transcription (Isuruta I. et al. 1998). Although the signaling pathway through AC/cAMP/PKA can negatively modulate the activity of transcriptional factors in directing the complex hypertrophic reprogramming, its activation appears to be necessary to enhance contractility and express the full hypertrophic phenotype. Our data suggest that, in the myocardium, calcineurin initiated transcriptional response leads to an enhanced AC5/6 protein expression which may be indicative of different patterns of genetic reprogramming of cAMP signaling in cardiac hypertrophic remodeling versus non-excitable cells. It is of interest that remodeling initiated by calcineurin did not appear to affect Ca^{2+}-dependent contractile reserve, but might have predisposed the hearts to a general decline in responsiveness to receptor systems which operate at the level of PKC and βARK1 (Winstel R. et al. 1996, Lemire I. et al. 1998) . Up-regulation of Ca^{2+}-dependent PKC isoform(s) intrinsic to the myocardium with chronically activated calcineurin (De Windt L. et al. 2000) can cause desensitization of the β-AR through activation of βARK1. Thus, PKC-mediated functional uncoupling may affect upstream signaling and may be responsible to some extent for the observed blunting or uncoupling of the β-AR signaling pathway in calcineurin-mediated hypertrophy. Direct stimulation of AC activity with forskolin also produced a reduced inotropic response, suggesting that cAMP production is significantly affected in the transgenic model under conditions of physiological extracellular calcium concentration and enhanced contractility. We achieved a restoration of the contractile response to β-AR stimulation by lowering the calcium in the perfusing solution and therefore diminishing calcium influx through L-type voltage-dependent calcium channels. The prevalence of AC5 and AC6 in the cardiomyocytes hints at a crucial role for the susceptibility of cAMP-dependent signaling to calcium inhibition (Yo H.J.et al. 1993). Up-regulation of the cellular signaling pathway through AC5/6 found in the transgenic heart with calcineurin overexpression may represent a part of maladaptive changes and in addition serving as a mechanism limiting cAMP production through stimulated β-adrenoreceptors.

Our data indicated that Ca^{2+}-AC interaction independent from calcineurin is the most dynamic target affected in the hypertrophied heart with blunted β-AR responsiveness. Increased calcium influx and up-regulated AC5/6 may represent a transcriptional mechanism by which this regulatory cascade acquires calcium to inhibit cAMP production. The calmodulin/calcineurin/AC9 pathway can be ruled out as a possible potential target for increasing cardiac responsiveness to adrenergic stimulation, and targeting Ca^{2+}-dependent AC isoforms may represent an additional approach to manipulate the effectiveness of the β-adrenergic pathway in cardiac hypertrophy and failure.

Prolongation of the action potential (AP) is well documented by many to be a prominent feature of the heart in hypertrophy or cardiac failure in a variety of species, involving altered expression of depolarizing and repolarizing currents (Wickenden A. et al. 1998*)*.

Calcineurin also has an important role in cardiac electrophysiologic remodeling as an opposing effector of kinase actions involving transcriptional as well as non-transcriptional mechanisms that can affect cardiac function. In contrast to other models of cardiac hypertrophy that show a significant downregulation of I_{to} (the transient outward potassium current) and I_{K1} (Mitarai S. et al. 2000, Knollmann B. et al. 2000, Oudit G.Y. et al. 2001), in the calcineurin-overexpressed hypertrophic model, we observed a significant increase in

$I_{to,peak}$ and I_{sus} but no significant change in I_{to} Although there is evidence that $I_{to,fast}$ is a key contributor in shaping the early phase of cardiac ventricular action potential, $I_{to,s}$ and the delayed rectifier currents (I_{Kur} and $I_{K,slow}$) also play a major role in repolarization of the mouse myocytes. According to model predictions, reduction in both $I_{to,fast}$ and I_{K1} current densities measured in terminal heart failure have only modest effects on the AP duration and the AP prolongation is primarily due to altered expression of intracellular Ca^{2+}-handling proteins, our observations suggest that the attenuation of the Kv1.5 and Kv2.1 α-subunits and the increased I_{Ca} may explain the marked action potential prolongation in transgenic cardiomyocytes. Previous electrophysiological recordings from cardiomyocytes isolated from transgenic mice expressing a mutant Kv2.1 α-subunit (Kv2.1 N216), which functions as a dominant negative, revealed that $I_{K,slow}$ was selectively attenuated and led to marked increases in action potential duration. The observed changes in $I_{to,peak}$ and I_{sus} in calcineurin-overexpressing transgenic mouse heart should in fact contribute to a *shortening* of the action potential duration. (Petrashevskaya N. et al. 2002). Therefore, it remains conceivable that the increased $I_{to,peak}$ may provide a 'compensatory' response to prevent excessive lengthening of the action potential duration and Ca^{2+} influx through LVDCC during excitation (Li Q. et al.1994). A new electrophysiological profile in myocardium with chronically activated calcineurin can contribute to the enhanced calcium influx, leading to a general decline in neurohumoral responsiveness. In addition, the augmented calcium influx through LVDCC acts as a trigger for SR Ca^{2+} release and there is evidence of an interaction between calcineurin and RYR2 increasing SR calcium loading (Valdivia H. 1998, Shou W. et al.1998).

6. CALCIUM CHANNEL BLOCKERS AS A TREATMENT FOR HEART FAILURE

As we discussed above, persistently increased Ca^{2+} influx through Cav1.2 and increases in the calcium transient amplitude and diastolic Ca^{2+} at high contracting rates activate the signaling cascades to stimulate pathologic hypertrophy. Calcium channel inhibition represents a logical approach to treatment of left ventricular hypertrophy. Based on the chemical structure, calcium channel blockers CCBs are categorized into 3 subroups; benzodiazepines (BTZs)(diltiazem, clenazem), phenylanalkylamines (PAAs) (verapamyl and gallopamin) and dihydropyridines (DHPs) (nifedipine, ilnidipine, felonidipine, manidipine, amlopidine, aranidipine, azelnidipine, cilnidipine, efonidipine, and nilvadipine) The differences in chemical structures would provide heterogeneity in the action of the agents. CCBs binds specifically to regions of the ά1-subunit of the L type of VDCC. Several laboratories have identified individual amino acids within the L-VDCC ά1 subunit that participate in the formation of the major drug-binding domains. The 3 classes of organic CCBs has separate , but overlapping or allosterically linked, Ca^{2+} channel-binding sites at IIIS6 and IVS6, IVS6 and IIIS6 and IVS6 motifs (He M. et al. 1997).

The importance of the voltage –dependent L-type calcium channels has been confirmed in many animals' models that demonstrate reduction in hypertrophy by calcium channels blockers, but there are limited data on the effect of L-type calcium channel blockade on cardiac hypertrophy in humans beyond blood pressure control. Clinically, calcium channel antagonists decrease blood pressure and induce regression of left ventricular hypertrophy, but prolongation of survival with these agents has not been demonstrated (Kingbeil A. et al. 2003).

Framingham study reported that the heart rate is one of the most important factors for the mortality of cardiovascular disease. CASTEL study also reported that heart rate is the strong predictor of the cardiovascular death in men (Palatine P. et al. 1999). It is conventionally believed that CCB were not an appropriate treatment for heart disease mainly because they elicit the reflex tachycardia associated with excessive decrease in blood pressure, activation of sympathetic nervous system and rennin-angiotensin system, cardiac overload and labile hypertension (Ishikawa K. et al. 1995). Blockade of non-cardiac channels explains many of the undesired effect such as systemic hypotension, constipation, edema that could also account for the dubious benefit on cardiovascular mortality. These drugs are not currently judged as being effective in the setting of congestive heart failure. Most of these problems are associated with the use of short acting dihydropyridines and the subsequent occurrence of reflex tachycardia and rapid blood pressure decrease. To counter this confounding issue, novel types of CCBs have been developed. All clinical trials to date, with the exception of the Prospective Randomized Amlodipine Survival Evaluation (PRAISE I) revealed that improved clinical symptoms were seen only in patients with heart failure of non-ischemic cardiomyopathic nature. The CCBs with sustained action to avoid reflex tachycardia such as amlodipine, lercanidipine and azelnidipine are included in this class. In contrast, chronic nifedipine with strong peripheral vasodilating effect caused a high incidence of clinical deterioration and worsening of heart failure (Eikayam U. at al. 1990, Furberg C. et al. 1950). Felodipine, administered to patients with congestive heart failure in a settings of stable therapy with enalapril, diuretics and digoxin, has neither a beneficial nor deteriorating effect despite the improvement in exercise performance and LV function (Dunselman P. 1989, Boden W. at all. 1996).

Gene therapy allows for more directed and organ specific delivery, and thus may avoid undesired systemic effects, which may account in part for the undesired clinical outcome with these agents. Genetic calcium channel blockade can be achieved in the heart, by overexpressing the small G-protein, GGem, by adenoviral gene transfer (Murata M. et al., 2004). Cingolani with coauthors (Cingolani E.et al. 2007) incorporated a short harpin RNA (shRNA) template sequence capable of mediating the knockdown of the L-type calcium channel accessory β-subunit gene that resulted in partial inhibition of L-type calcium current. In neonatal cardiac myocytes, L-type calcium channel subunit gene knockdown reduced calcium transient amplitude. In cellular model of hypertrophy, down-regulation of voltage dependent L-type calcium channels β-subunit expression prevented an increase in relative cell size and abrogated phenylephrine-induced protein synthesis. In vivo gene transfer attenuated the hypertrophic response in an aortic-banded rat model of left ventricular hypertrophy, with reduced left ventricular wall thickness and heart weight/body weight ratios. Genetic suppression of the L-type calcium channel accessory β-subunit with shRNA modulated calcium current and suppressed cardiac hypertrophy. The attenuated hypertrophic response in the setting of aortic constriction without difference in systemic blood pressure between the non-silencing and β -subunit suppressive groups, support a direct cardiac effect of voltage-dependent calcium channel β-subunit knockdown independent of changes in peripheral vascular resistance. There was not a depressed cardiac contractility and systolic performance from voltage dependent calcium channels gene silencing. These results are in accordance with finding with endogenous β-subunit trapping reported by Serikov et al (Serikov V. et al. 2002).

Focal modulation of voltage–dependent calcium channel β-subunit by a vector, capable of chronically suppressing gene expression represents an attractive alternative therapy in hypertrophic cardiomyopathy. Further mechanistic studies are required to elucidate any potential antihypertrophic effects of voltage –dependent calcium channels β-subunit knockdowns that are independent of the effects on calcium currents

7. Targeting Presynaptic Mechanisms of Catecholamines Release as a Rescue for Cardiac Hypertrophy

Despite resent advancement in prevention and management of heart disease, death due to chronic heart failure continues to rise and new and innovative treatments are needed. A salient feature of HF is elevated sympathetic nervous system (SNS) activity and outflow, reflected by increased circulating catecholamines. Initially an adaptive process to compensate for decreased function following cardiac injury through stimulation of β-AR receptors, SNS activation becomes maladaptive, contributing significantly to disease morbidity and mortality. Heart failure is a disease characterized not only by alterations in myofilaments, and extracellular matrix, defective systolic and diastolic calcium cycling and impaired ventricular function, but also by a loss of adaptive capabilities, causing such symptoms as exercise intolerance. Chronic adrenergic signaling early on is helpful but ultimately is harmful compensatory mechanism developed by the diseased human heart.

In chronic cardiac insufficiency, persistent low cardiac output stimulates continuous sympathetic activation and catecholamine release, which induces the uncoupling of myocardial β-adrenergic receptors, further diminishing cardiac pump performance. In the failing heart, β-adrenergic signal is reduced secondary to desensitization changes in β1 and possibly β2 receptors. Levels of norepinephrine (NE) are associated with worsened prognosis in HF. Heart failure from virtually every etiology is accompanied by enhanced sympathetic activity; an adaptation in response to decreases cardiac output. While this response is effective in increasing contractility during acute decompensation, prolonged activation is deleterious, leading to worsening failure (Chacraborti S et al 2000, Engelhardt S. et al. 2004, Petrashevskaya N. et al. 2008).

The basic knowledge about mechanisms by which sustained adrenergic activation promotes myocardial growth, as well as understanding how structural changes in hypertrophied myocardium could affect myocardial function has been acquired from studies using an animal model of chronic systemic beta-adrenoreceptor agonist administration. Sustained beta-adrenoreceptor activation was shown to enhance the synthesis of myocardial proteins, an effect mediated via stimulation of myocardial growth factors, up-regulation of nuclear proto-oncogenes, induction of cardiac oxidative stress, as well as activation of mitogen-activated protein kinases and phosphatidylinositol 3-kinase. Sustained beta-adrenoreceptor activation contributes to impaired cardiac autonomic regulation as evidenced by blunted parasympathetically-mediated cardiovascular reflexes as well as abnormal storage of myocardial catecholamines. Catecholamine-induced cardiac hypertrophy is associated with reduced contractile responses to adrenergic agonists, an effect attributed to down-regulation of myocardial beta-adrenoreceptors, uncoupling of β-adrenoreceptors and adenylate cyclase, as well as modifications of downstream cAMP-mediated signaling. In compensated cardiac hypertrophy, these changes are associated with preserved or even enhanced basal ventricular systolic function due to increased sarcoplasmic reticulum Ca^{2+} content and Ca^{2+}-induced sarcoplasmic reticulum Ca^{2+} release. The increased availability of Ca^{2+} to maintain cardiomyocyte contraction is attributed to prolongation of the action potential due to inhibition of the transient outward potassium current as well as stimulation of the reverse mode of the Na^{+}/Ca^{2+} exchange. Further progression of cardiac hypertrophy towards heart failure is due to abnormalities in Ca^{2+} handling, necrotic myocardial injury, and increased myocardial stiffness due to interstitial fibrosis (Petrashevskaya N. et al. 2008, Hiroyuki Nakayama et al. 2007).

Cardiac hypertrophy has been shown to enhance ET-1/NGF (myocyte derived nerve – growth factor) signaling that causes sympathetic hyper-innervation. However, the anatomical sympathetic hyper-innervation occurs simultaneously with depressed neuronal function. It is possible that negative feed-back mechanisms might underlie the down-regulation of neuronal function to prevent excess signaling from the hyper-innervated sympathetic nerves (Kensuke Kimura ET al.2007).

Hemodynamic derangements and neurohumoral activation have been hallmarks of heart failure. Standard pharmacological β-adrenergic receptor (β -AR) antagonists prolong survival in heart failure at the expense of β-AR unresponsiveness and the possibility of exacerbating heart failure. Coordinated attack on excessive sympathetic activation includes novel pharmacotherapies targeting alpha-2 receptor mediated feedback inhibition of catecholamine secretion in adrenal glands (Hein L.et al. 1999), central and peripheral neurohormonal pathways targeting aldosterone and AT1 receptor blockade (Huang B. et al. 2002, Lal A. et al. 2004).

Among 3 types of VDCCs, P/Q –type, and R-type, are predominantly expressed in neuronal cells. N- type Ca^{2+} calcium channels have been recognized as high-voltage-activated Ca^{2+} channels selectively sensitive to blockade by ω-conotoxin, a 27 –aminoacid peptide isolated from the venom of the fish-hunting cone snail Conus geographus, but resistant to the L-type Ca^{2+} channels specific blockers such as dihydropyridines. Molecular studies have revealed that α1b (Cav2.2) subunit genes encodes N type Ca^{2+} channels and is expressed in neuronal tissues (Ino M. et al. 2001), peripheral neurons such as autonomic neurons and motor neurons. N –type calcium channels have been shown to be critically involved in the release of neurotransmitters from nerve terminals. Selective elimination of N-type calcium channels in sympathetic nerve terminals in ά1b-deficient mice resulted in markedly diminished sympathetic nerve activity but parasympathetic nerve activity was unaffected. It is possible that other types of VDCCs can substitute for the N-type Ca^{2+} channels in inducing certain neuronal processes, in contract to the essential role played by the N- type Ca^{2+} channels in sympathetic nerve activity.

Considerable importance to counteract the overactive neurohormonal system has shifted attention to N-type voltage –dependent calcium channels (N-type VDCC) ($Ca_v2.2$) channel blockers acting to limit the hyper-activation of catecholaminergic signaling. A novel approach in drug development and application to limiting sympathetic drive is to interrupt the presynaptic neurotransmitter release and minimize sympathetic reflex during antihypertensive therapy. The N-type calcium channels in the presynaptic nervous ending are targets for calcium channel antagonists that act to reduce stimulus- release efficiency and subsequently catecholamine secretion and essential sympathetic activation. N-type VDCC blockers have been proven to have benefits of decreasing baroreflex-induced tachycardia through prevention of sympathetic reflex activation after a decrease in blood pressure, which is particularly important in   ischemic heart disease. Experimental and clinical evidences exist that the suppression of sympathetic nerve activity by the blockade of N-type calcium channels contributes to the improvement of LV hypertrophy and diastolic function (Takami T. et al. 2003). Cilnidipine is a calcium channel blocker, whose action lasts longer than that of the first generation of Ca^{2+} channels blockers including nifedipine, and the drug is used for the treatment of essential hypertension. Cilpinidine is expected to be a favorable drug that overcomes some of the deficiencies of the first generation of Ca^{2+} channel blockers. N- type calcium channel blocker ω-conotoxin suppressed the sympathetic nerve stimulation-induced pressor response but did not affect the angiotension II-induced one. Cilnidipine in antihypertensive doses suppressed the sympathetic verve stimulation-induced pressor response more potently that the angiotensin II-induced one, in contract to nifedipine, suggesting that cilnidipine can suppress norepinephrine release from sympathetic nerve endings through the activation of N-type Ca^{2+} channels in addition to vascular contraction through activation of L-type Ca^{2+} channels on vascular smooth muscle.

The study comparing the effects of amlodipine, benidipine and nifedipine on myocardial hypertrophy revealed that that in myocardial hypertrophy model created by transverse aortic constriction (TAC) in C57 BL/6 mice, plasma catecholamine concentrations was significantly increased in TAC mice 1 week after surgery in comparison to sham operated rats. Treatment with amlodipine or benidipine, but not nifedipine, decreased such hypertrophic marker as heart weight/body weight ratio (Luo Q. et al. 2010). Anti-hypertrophic effect of calcium channel blockers is dependent on their potential of blocking N-type calcium channel, and the underlying mechanism involves the sympathetic inhibition. Hypertensive patients characteristically exhibit left ventricular hypertrophy and diastolic dysfunction. The effect of hypotensive agents on LV hypertrophy and diastolic dysfunction do not always correlate with the degree of blood pressure reduction, but their effects on the sympathetic nervous system are thought to be important.

To establish a window of opportunities for therapeutic interventions, targeting N-type VDCC, it is necessary to understand the consequences of long-term N-type calcium channel blockade on cardiac function and regulation and limitations of N–type calcium channel blockade. One important challenge is that effectiveness of N-type calcium channels antagonists may be reduced by compensatory adjustment in presynaptic sympathetic coupling -release. The increased subunit interaction between the auxiliary β3 subunit and the R-type calcium channel (Cav2.3) α1 subunit has been reported to contribute to maintained catecholamine release in the absence of N-type channels. These presynaptic adjustments may reduce the effectiveness of sympatholitic therapy via N-type calcium channel inhibition. One of the adverse long-term consequences of N-type calcium channel blockade that may limit use of N-type calcium channels blockers is an increased α_1-adrenergic vasoreactivity and, consequently, elevated mean arterial blood pressure. Importantly, the knock-out mice showed elevated basal mAP and HR despite sympathetic nerve dysfunction. Elevated mAP and HR in mutant mice are not attributable to elevated noradrenalin in plasma or to up-regulation of β-adrenergic receptors, as seen in sympathetic denervation. Contraction mediated by ά2-adrenergic receptors or thromboxane A2 receptors was potentiated in the isolated thoracic aorta from N-type-deficient mice. Future studies are necessary to reveal genes whose expression is deregulated via presynaptic N-type channels to induce abnormalities in the postsynaptic cardiovascular tissues (Mori Y. et al. 2002).

Diminished sympathetic inflow from presynaptic endings can enhance postsynaptic myocardial responsiveness to sympathetic agonists. We demonstrated recently that elective impairment of sympathetic inflow does not modulate postsynaptic β-adrenoreceptor (β –AR) expression and coupling, but causes increased functional response to secondary system of inotropic regulation via pro-growth Gαq-protein coupled α_1-adrenergic receptors(α_1. AR) (Petrashevskaya N. et al. 2010)

The effect of α_1-AR stimulation on cardiac contraction induces positive and negative inotropy. The molecular components for evoking effects on cardiac contractility occur via different transduction pathways initiated from α_{1A}- and α_{1B}-AR. The cardiac α_1-AR signaling pathway diverges at the level of the α_1-AR subtype and the G-proteins, which produce the opposite effect on I_{Ca} in rat ventricular myocytes. α_{1A}-AR are coupled with Gq/1 and activate the PLC-PKC-CaMKII pathway evoking the potentiation of I_{Ca}. The α_{1B}-AR interacts with Gi, of which the $\beta\gamma$-complex couples directly to inhibit I_{Ca} (O-Uchi J. et al. 2008). In N type-/- mice, increased α_1-AR receptor mediated contractile responses might be related to increased phospholipase Cβ_1 (PLCβ_1) expression. Stimulation of α_1-AR has been shown to selectively generate activation of PLCβ_1, in contrast to PLCβ_3, another cardiac isoform, PLCβ1 has been suggested to be a major mediator of Go-initiated hypertrophic growth of the myocardium. PLCβ1 may be a candidate for the sensor of reduced presynaptic sympathetic transmission in the heart regulated by sympathetic nerve activity via presynaptic N-type channels. Activated PLCβ1-PKC-CaMKII could directly potentiate I_{Ca} through the phosphorylation of the α_1 and/or β subunits, increasing calcium influx and compensating for the diminished presynaptic norepinephrine release. Serving as a mechanism of secondary contractile support, Gαq/PLCβ_1 represents the signaling pathway for endothelin I, angiotensin II, and α_1-AR receptors. Activation of α_{1A}-AR selectively activates the PKC-CaMKII leading to increased calcium influx through L-type VDCCs. The molecular mechanisms of this functional response on cardiac contractility was related to increased expression of PLCβ1 specifically coupled to α_1-ADR-Gq pathway. It is possible that PLCβ1 may be a candidate for the sensor of reduced catecholamine release regulated by sympathetic nerve activity via presynaptic N-type calcium (Petrashevskaya N. et al. 2011). This activation is implicated in the development of pressure overload hypertrophy by activating some canonical transient receptor potential channel. Thus, enhanced responsiveness of this pro-hypertrophic cascade may predispose the myocardium to develop hypertrophic remodeling after pressure overload.

Taking together, our findings confirm that the functional redundancy of VDCC's prevents aberrant neurotransmission in parasympathetic nervous endings while sympathetic transmission is much more susceptible to the lack of calcium influx through N-type calcium channels. It also places into question whether this partial lack of sympathetic transmission has no adverse long-term effect on balance between growth promoting and growth inhibiting cascades.

CONCLUSION

In this review, we examined new evolving heart failure therapies based on N-type and L-type calcium channel blockers and Ca-dependent signaling pathways modulators. These new therapeutics target sympathetic hyper-activation and Ca-dependent proximal signaling with intent to prevent, halt, or possible reverse the progression of disease.

Considerable importance to counteract the overactive neurohormonal system has shifted attention to N-type voltage –dependent calcium channels (N-type VDCC) (Cav2.2) blockers to limit the hyper-activation of catecholaminergic signaling. N-type VDCC blockers have been proven to have benefits of decreasing baroreflex-induced tachycardia through prevention of sympathetic reflex activation after a decrease in blood pressure, which is particularly important in ischemic heart disease. Experimental and clinical evidences exist that the suppression of sympathetic nerve activity by the blockade of N-type calcium channels contributes to the improvement of LV hypertrophy and diastolic function. To establish a window of opportunities for therapeutic interventions, targeting N-type VDCC, it is necessary to understand the consequences of long-term N-type calcium channel blockade on cardiac function and regulation and limitations of N –type calcium channel blockade.

The L type VDCC (Cav1.2) are an essential part of an integrated calcium management in cardiac hypertrophy and heart failure for regulating not only cardiac inotropy but also gene expression governing cardiac remodeling. Changes in L-type calcium channels activity (both frequency – and amplitude dependent) influence a variety of signaling pathways, regulate normal cardiac metabolism, and cause both physiological and pathological hypertrophy. Although the central role of altered Ca^{2+} signaling is implicated in hypertrophic signaling pathways, the specific role of Ca^{2+} influx through the L-type Ca2+ channel and the efficacy of L-type calcium blockers in cardiac hypertrophy and failure have been controversial. Effects of these drugs have not emerged as unequivocally favorable in all clinical studies to date. The functional and structural Ca^{2+} channel remodeling also occurs in human heart failure and by it provoke pathophysiological consequences on cardiac morphology, contraction, or rhythm. Molecular and functional analyses have shown that in heart failure enhanced L-VDCC activity is accompanied by altered expression pattern of auxiliary L-VDCC beta-subunit gene products, especially β2 subunit .These data support the concept that in human heart failure, alterations of auxiliary subunit play an important role and at least some types of heart failure can be considered as L-type VDCC β subunit channelopathies.

A growing body of studies now reveals that N- and L-type VDCC play an important role in the pathogenesis of cardiac hypertrophy and congestive heart failure. The therapeutic effectiveness of calcium channel antagonists can be enhanced by targeting multiple channels subtypes, including L-, N-type and their auxiliary subunits.

REFERENCES

Altier C, Garcia-Caballero A, Simms B, You H, Walcher J, Tedford H, Hermosilla T, Zamponi G. The Cavβ subunit prevents RFP2-mediated ubiquitination and proteasomal degradation of L-type channels. Nat Neurosci. 2011; 14:173-80.

Antoni F, Palkovits M, Simpson J, Smith S, Leitch A, Rosie R, Fink G, Paterson Jl. Ca2+/calcineurin-inhibited adenylyl cyclase, highly abundant in forebrain regions is important for learning and memory. J Neurosci. 1998;18:9650-966.

Bare D, Kettlun CS, Liang M, Bers D, Mignery G. Cardiac type inositol 1,4,5-triphosphate receptor: interaction and modulation by calcium/calmodulin dependent protein kinase II J. Biol Chem. 2005;280:15912-15920.

Beetz N, Lutz H, Meszaros J, Gilsbach R, Barreto F, Meissner M, Hoppe U, Schwartz A, Herzig S, Matthes J. Transgenic simulation of human heart failure-like L-type Ca2+-channels: implications for fibrosis and heart rate in mice. Cardiovasc Res. 2009;84:396-406.

Belin R, Sumandea M, Allen E, Schoenfelt K, Wang H, Solaro R, deTombe P. Augmented protein kinase C-alpha-induced myofilament protein phosphorylation contributes to myofilament dysfunction in experimental congestive heart failure. Circ Res. 2007;101;195-204.

Bers D. Calcium fluxes involved in control of cardiac myocyte contraction. Circ Res. 2000;87:275-281.

Berns D, Guo T. Calcium signaling in cardiac ventricular myocytes. Ann.N.Y.Acad.Sci 2005;1047:86-98.

Boden W, Ziesche S, Carson P, Conrad C, Syat D, Cohn J. Rationale and design of the Third Vasodilator-Heart Failure Trial (V-HeFTIII):felodipine as adjunctive therapy to enalapril and loop diuretics with or without digoxin in chronic congestive heart failure V-HeFTIII investigators. Am J. Cardiol. 1996;77:1078-1082.

Bodi I, Mikala G, Koch SE, Akhter SA, Schwartz A. The L-type calcium channel in the heart: the beat goes on. J Clin Invest. 2005;115:3306–17.

Bodor J, Habener F. Role of transcriptional repressor ICER in cyclic AMP-mediated attenuation of cytokine gene expression in human thymocytes. J Biol Chem. 1998;273:9544-9551.

Bouzamondo A, Hulot J, Sanchez P, Cucherat M, Lechat P. Beta-blocker treatment in heart failure. Fundam. Clin. Pharmacol. 2001;15:95–109.

Bowling N, Walsh R, Song G, Estrtridge T, Sandusky G, Fouts R, Mintze K, Pickard T, Roden R, Bristow M, Sabbah H, Mizrahi J, Gromo G, King G, Vlahos C. Increased protein kinase C activity and expression of Ca2+-sensitive isoforms in the failing human heart. Circulation. 1999;99:384 –391.

Braz J, Gregory K, Pathak A, Zhao W, Sahin B, Klevitsky R, Kimball T, Lorenz J, Nairm A, Liggett S, Bodi I, Wang S, Schwartz A, Lakatta E, DePaoli-Roach A, Robbins J, Hewett T, Bibb J, Westfall M, Kranias E, Molkentin J. PKCa regulates cardiac contractility and propensity towards heart failure. Nature Med. 2004;10:248-254.

Bristow M.R. Beta-adrenergic receptor blockade in chronic heart failure. Circulation. 2000;101:558–569.

Carafoli E, Santelia L, Branca D, Brini M. Generation, control, and processing of cellular calcium signal. Crit Rev Biochem Mol Biol . 2001;36:107-260.

Chakraborti S, Hakraborti T, Shaw G. β-adrenergic mechanisms in cardiac diseases: a perspective. Cell Signal. 2000;12:499-513.

Chen X, Piacentino V, Furukawa S, Goldman B, Margulies KB, Houser SR. L-type Ca2+ density and regulation are altered in failing human ventricular myocytes and recover after support with mechanical assist devices. Circ.Res. 2002;91:517-524.

Chiang C, Huang C, Chieng H., Chang Y., Chang D., Chen J . The Cav3.2 T type Ca2+ channel is required for pressure-overload-induced cardiac hypertrophy in mice. Circ Res 2009;65:522-530.

Cho M, Rapacciuolo A, Koch W, Kobayashi Y, Jones L, Rockman H. Defective-adrenergic receptor signaling precedes the development of dilated cardiomyopathy in transgenic mice with calsequestrin overexpression. J Biol Chem. 1999;274:22251–22256.

Cingolani E, Ramirez Correa G, Kizana E, Murata M, Cheol Cho H, Marban E. Gene therapy to inhibit the calcium channel b-subunit:physiological consequences and pathophysiological effects in models of cardiac hypertrophy. Circ Res. 2007;101:166-175.

Colecraft H, Alseikhan B, Takahashi S, Chaudhuri D, Mittman S. Novel functional properties of Ca(2+) channel beta subunits revealed by their expression in adult rat heart cells. J Physiol. 2002;541:435–452.

Colella M, Grisan F, Robert V, Turner V, Thomas A, Pozzan T. Ca2+ oscillation frequency decoding in cardiac hypertrophy: Role of calcineurin/NFAT as Ca2+ signal integrators. PNAS. 2008;105:2859-2864.

Colucci W, Sawyer D, Singh K, Communal C. Adrenergic overload and apoptosis in heart failure: implications for therapy. J. Card. Fail. 2000;6:1–7.

Cooper D, Mons N, Karpen J. Adenylyl cyclases and the interaction between calcium and cAMP signalling. Nature. 1995; 374: 421 -424.

Crompton M, Barksby E, Johnson N, Capano M. Mitochondrial intermembrane junctional complexes and their involvement in cell death. Biochimie. 2002;84:143–152.

Danial N, Korsmeyer S, Cell death: critical control points. Cell. 2004;116:205–219.

Zamzami N, Kroemer G. The mitochondrion in apoptosis: how Pandora's box opens. Nat. Rev. Mol. Cell Biol. 2001;2:67–71.

Davies A, Hendrich J, Van Minh A, Wratten J, Douglas L, Dolphin A. Functional biology of the alpha(2)delta subunits of voltage-gated calcium channels. Trends Pharmacol Sci. 2007;28:220–228.

De Windt L, Hae W, Haq S, Force T, Molkentin J. Calcineurin promotes protein kinase C and c-Jun NH2-terminal kinase activation in the heart. J Biol Chem. 2000;275:13571-13579.

Dorn G., Force T. Protein kinase cascades in the regulation of cardiac hypertrophy. J Clin Invest. 2005;115:527-537.

Dunselman P. Efficacy of felodipine in congestive heart failure. Eur Heart.J. 1989;10:354-364.

Eisner D., Trafford A. No role for the ryanodine receptor in regulating cardiac contraction? News Physiol. 2000;15:275-279.

Elkayam U, Amin J, Mehra A, Vasquez J, Weber L, Rahimtoola S. A prospective, randomized, double-blind, crossover study to compare the efficacy and safety of nifedipine therapy with that of isosorbide dinitrate and their combination in treatment of chronic congestive heart failure. Circulation. 1990;82:1954-1961.

Engelhardt S, Hein L, Dyachenkow V, Kranias E, Isenberg G, Lohse M. Altered calcium handling is critically involved in the cardiotoxic effects of chronic beta-adrenergic stimulation. Circulation. 2004;109:1154–1160.

Frace A, Hartzell H. Opposite effects of phosphatase inhibitors on L-type calcium and delayed rectifier currents in frog cardiac myocytes. J Physiol (Lond) 1993;472:305-326.

Fuller M, Emrick M, Sadilek M, Scheuer T, Catterall W. Molecular mechanism of calcium channel regulation in the fight-or-flight response. Sci Signal. 2010; 3:70

Foo R, Mani K, Kitsis R. Death begets failure in the heart. J Clin Invest. 2005;115:565–571.

Furberg C, Psary B, Meyer J. Nifedipine. Dose-related increase in mortality in patients with coronary heart disease. Circulation. 1995;92:1326-1331.

Gao M, Ping P, Post S, Insel P, Tang R, Hammond H. Increased expression of adenylylcyclase type VI proportionately increases beta-adrenergic receptor-stimulated production of cAMP in neonatal rat cardiac myocytes. Proc Natl Acad Sci USA. 1998;95:1038-1043.

Gao M, Lai N., Roth D., Zhou J, Zhu J, Anzai T, Dalton N, Hammond H . Adenylylcyclase increases responsiveness to catecholamine stimulation in transgenic mice. Circulation. 1999;99:1618-1623.

Go L, Moshella M, Watras J., Handa K, Fyfe B, Marks A. Differential regulation of two types of intracellular calcium release channels during end-stage heart failure. J Clin Invest. 1995;95:888-894.

Grimm M, Brown J. B-Adrenergic receptor signaling in the heart: Role of CaMKII. J Mol Cell Cardiol. 2010;48:322-330.

Groner F, Rubio M, Schulte-Euler P, Matthes J, Khan I, Bodi I, Koch S, Scwartz A, Herzig S. Single-channel gating and regulation of human L-type calcium channels in cardiomyocytes of transgenic mice. Biochem Biophys Res Commun. 2004;314:878–884.

Haase H, Kresse A, Hohaus A, Schulte H, Maier M. Expression of calcium channel subunits in the normal and diseased human myocardium. J Mol Med. 1996;74:99–104.

Halestrap A. Calcium, mitochondria and reperfusion injury: a pore way to die. Biochem. Soc. Trans. 2006;34:232–237.

Hambleton M, Hahn H, Pleger S, Kuhn M, Klevitsky R, Carr A, Kimball T, Hewett T, Dorn G, Koch W, Molkentin J. Pharmacologic- and gene therapeutic-based inhibition of PKCa/b enhances cardiac contractility and attenuates heart failure. Circulation. 2006;114:74-582.

Haq S, Choukroun G, Lim H. Differential activation of signal transduction pathways in human hearts with hypertrophy versus advanced heart failure. Circulation. 2001;103:670-677.

Haldeman G, Croft J, Giles W, Rashidee A. Hospitalization of patients with heart failure: National Hospital Discharge Survey, 1985 to 1995. Am. Heart. J. 1999;137:352–360.

He M, Bodi I, Mikala G, Schwartz A. Motif III S5 of L-type calcium channels is involved in the dihydropyridine binding site. A combined radioligand binding and electrophysiological study. J. Biol.Chem. 1997;272:2629-2633.

Hein L, Altman J, Kobilka B. Two functionally distinct alpha2-adrenergic receptors regulate sympathetic neurotransmission. Nature. 1999.402:181-184.

Hobbs R. Guidelines for the diagnosis and management of heart failure. Am. J. Ther. 2004;11:467–472.

Heineke J, Molkentin J. Regulation of cardiac hypertrophy by intracellular signaling pathways. Nat Rev Mol Cell Biol. 2006;7:589-600.

Hoppe U, Marban E, Johns D. Adenovirus-mediated inducible gene expression in vivo by a hybrid ecdysone receptor. Mol Ther. 2000;1:159–164.

Houser S, Molkentin J. Does contractile Ca2+ control calcineurin-NFAT signaling and pathologic hypertrophy in cardiac myocytes. Sci Signal . 2008;24:pe31.

Houser S, Piacentino V, Weisser J. Abnormalities of calcium cycling in the hypertrophied and failing heart. J Mol Cell Cardiol. 2000;32:1595–1607.

Huang B, Ahmad M, Tan J, Leenen F. Sympathetic hyperactivity and cardiac dysfunction post-MI: different impact of specific CNS versus general AT1 receptor blockade J. Mol. Cell. Cardiol. 2002;43:479-486.

Huang L, Wolska B, Montgomery D, Burkart E, Buttrick P, Solaro R. Increased contractility and altered Ca(2+) transients of mouse heart myocytes conditionally expressing PKCbeta A. J Physiol Cell Physiol 2001;280:C1114-1120.

Hulme J, Yarov-Yarovoy V, Lin T, Scheuer T, Catterall W. Autoinhibitorycontrol of the channels by its proteolytically processed distal C-terminal domain. J Physiol .2006;576(Pt 1):87–102.

Hullin R, Asmus F, Ludwig A, Hersel J, Boekstegers P. Subunit expression of the cardiac L-type calcium channel is differentially regulated in diastolic heart failure of the cardiac allograft. Circulation. 1999;100:155–163.

Hullin R, Khan I, Wirtz S, Mohacsi P, Varadi G, Schwartz A, Herzig S. Cardiac L-type calcium channel beta-subunits expressed in human heart have differential effects on single channel characteristics. J Biol Chem. 2003;278:21623–21630.

Hullin R, Matthes J, von Vietinghoff S, Bodi I, Rubio M, D'Souza K, et al. Increased expression of the auxiliary beta2-subunit of ventricular L-type Ca2+ channels leads to single-channel activity characteristic of heart failure. PLoS ONE. 2007;2:e292

Hiroyuki Nakayama, Chen X, Baines A, Klevitsky R, Xiaoying Zhang X, Hongyu Zhang H., Naser Jaleel N, Balvin H, Chua T, Hewett T, Robbins J. Houser S, Molkentin J. Ca2+- and mitochondrial-dependent cardiomyocyte necrosis as a primary mediator of heart failure. J Clin Invest. 2007;117:2431-2444.

Ikeda K, Tojo K, Tokudome G, Akashi T, Hosoya T, Harada M et al Possible involvement of endothelin-1 in cardioprotective effects of benidipine. Hypertens Res 2000;23:491-495.

Ino M, Yoshinaga T, Wakamori M, Miyamoto N, Takahashi E, Sonoda J, Kagaya T, Oki T, Nagasu T, Nishizawa Y, Tanaka I, Imoto K, Aizawa S, Koch S, Schwartz A, Niidome T, Sawada K, Mori Y. Functional disorders of the sympathetic nervous system in mice lacking the alpha 1B subunit (Ca2.2) of N-type calcium channels. Proc Natl Acad Sci USA. 2001;98:5323-5328.

Ishikawa K, Nakai S, Takenaka T, Kanamasa K, Hama J, Ogawa I, Yamamoto T, Oyaizu M, Kimura A, Yamamoto K, Yabushita H, Katori R. Short acting nifedipine and diltiazemam do not reduce the incidence of cardiac events in patients with healed myocardial infarction. Secodary prevention group. Circulation. 1995:2368-2373.

Kajstura J, Cheng W, Sarangarajan R, Li B, Nitahara J, Chapnick S, Reiss K, Olivetti G, Anversa P. Necrotic and apoptotic myocyte cell death in the aging heart of Fischer 344 rats. Am. J. Physiol. 1996;271:H1215–H1228.

Kingbeil A, Scheider M, Martus P, Kamp T, Jia-Qiang He. L-type Ca2+ channels gaining respect in heart failure. Circ Res. 2002;91:451-453.

Kim J, Ghosh S, Nunziato D, Pitt G. Identification of the components controlling inactivation of voltage-gated Ca2+ channels. Neuron. 2004;41:745-754.

Kimura K, Ieda M, Kanazawa H, Yagi T, Tsunoda M, Ninomiya S, Kurosawa H, Yoshimi K, Mochizuki H, Yamazaki K, Ogawa S, Fukuda K. Cardiac sympathetic rejuvenation: a link between nerve function and cardiac hypertrophy Circ. Res. 2007; 100:1755-1764.

Knollmann B, Knollmann-Ritschel B, Weissman N, Jones L, Morad M. Remodelling of ionic currents in hypertrophied and failing hearts of transgenic mice overexpressing calsequestrin. J Physiol (Lond). 2000;525Pt2:483-498.

Koch W, Lefkowitz R., Rockman H. Functional consequences of altering myocardial adrenergic receptor signaling. Annu Rev Physiol. 2000;62:237-260.

Kobrinsky E, Abrahimi P, Duong S, Thomas S, Harry J, Patel S, Lao Q, Soldatov N. Effect of Cav? subunits on structural organization of Cav1.2 channels. PLoS ONE. e5587; 2009.

Lal A, Veinot J, Leenen F. Critical effect of aldosterone in cardiac remodeling post-myocardial infarction in rats. Cardiovasc. Res. 2004;64:437-447.

Lao Q, Kobrinsky E, Harry J, Ravindran A, Soldatov N. New determinant for the Cavbeta 2-subunit modulation of the Cav1.2 calcium channel. J Biol Chem. 283: 15577-15588, 2008

Lao Q, Kobrinsky E, Liu Z, Soldatov N. Oligomerization of Cavβ subunits is an essential correlate of Ca2+ channel activity. FASEB J. 2010;24: 5013-5023.

Levy D, Kencchaiah , Larson M, Benjamin E, Kupka M, Ho K, Murabito J, Vasan R. Long-term trends in the incidence of and survival with heart failure. N. Engl. J. Med. 2002;347:1397–1402.

Liu X, Yang P, Yang W, Yue D. Enzyme-inhibitor-like tuning of Ca(2+) channel connectivity with calmodulin. Nature. 2010; 463:968-72.

Liu Q, Chen X, MacDonnel S, Kranias E, Lorenz J, Leitges M, Houser S, Molkentin J. PPKCa, but not PKCβ or PKCγ, regulated contractility and heart failure susceptibility: implication for ruboxistaurin as a novel therapeutic approach. Circ Res. 2009; 17:194-200.

Liu Q. Molkentin J. Protein kinase Cα as a heart failure therapeutic target. J. Mol. Cell. Cardiol . 2010 .Article in Press

Li L, Yee C, Beavo J. CD3- and CD28-dependent inhibition of PDE7 required for T cell activation. Science. 1999;283:848-8517.

Luo Q, Xuan W, Xi F, Liao Y, Kitakaze M. Antihypertrophic effect of dihydropyridines calcium channel blockers is dependent on their potential of blocking N-type calcium channel. Nan Fang Yi Ke Da Xue Xue Bao. (Chinese). 2010;30:755-759.

MacLennan D, Kranias E. Phospholamban: a crucial regulator of cardiac contractility. Nat. Rev. Mol. Cell Biol. 2003;4:566–577.

Mackenzie L, Bootman M, Laine M, Berridge M, Thuring J, Holmes A, Li W, Lipp P . The role of inositol 1, 4, 5 –triphosphate receptors in Ca2+ signalling and the generation of arrhythmias in rat atrial myocytes. J Physiol. 2002 ;541:395-4090.

Malek M. Health economics of heart failure. Heart. 1999;82(Suppl. 4):IV11–IV13.,

Marks A. Heart failure and sudden cardiac death: Causes and cures. 2005. Harvey Lect. 101, 39–57.

Meguro T., Hong C., Asai K., Takagi G, McKinsey TA, Olson EN, Vatner SF. Cyclosporine attenuates pressure-overload hypertrophy in mice while enhancing susceptibility to decompensation and heart failure. Circ Res. 1999;84:735-740.

Messerli F, Schmieder R. A meta-analysis of the effects of treatment on left ventricular mass in essential hypertrophy. Am J Med. 2003;115:41-46.

Mitarai S, Reed T, Yatani A. Changes in ionic currents and β-adrenergic receptor signaling in hypertrophied myocytes overexpressing Gαq . Am J Physiol. 2000;279:H139-H148.

Mori Y, Nishida M, Shimizu S, Ishii M, Yoshinaga M, Ino M, Sawada K, Niidome T. Ca2+ channel a1b subunit (Cav2.2) knockout mouse reveals a predominant role of N type channels in the sympathetic regulation of the circulatory system. Trends Cardiovasc Med. 2002;12:279-275.

Mukherjee R, Spinale F. L-type calcium channel abundance and function with cardiac hypertrophy and failure: a review. J Mol Cell Cardiol. 1998;30:1899-1916.

Muller FU, Neuman J, Schmitz W. Transcriptional regulation by cAMP in the heart. Mol Cell Biochem. 2000;212:11-17.

Muth J, Varadi G, Schwartz A. Use of transgenic mice to study the voltage-dependent Ca2+ calcium channels. TIPS.2001;22:526-532.

Muth J, Yamaguchi H, Mikala G, Grupp I, Lewis W, Cheng H, Song L, Lakatta E., Varadi G., Schwartz A. Cardiac-specific overexpression of the alpha(1) subunit of the L-type voltage-dependent Ca(2+) channel in transgenic mice. Loss of isoproterenol-induced contraction. J Biol Chem. 1999; 30;2:21503-6.

Muth J, Bodi I, Lewis W, Varadi G, Schwartz A.A Ca(2+)-dependent transgenic mouse model of cardiac hypertrophy: A role for proteine kinase Calpaha. Circulation. 2001;103:140-147.

Murakami M, Yamamura H, Suzuki T, Kang M, Ohya S, Murakami A, Miyoshi I, Sasano H, Muraki K, Hano T, Kasai N, Nakayama S, Campbell K, Flockerzi V, Imaizumi Y, Yanagisawa T, Iijima T. Modified cardiovascular L-type channels in mice lacking the voltage-dependent Ca2+ channel beta3 subunit. J Biol Chem. 2003;278:43261–43267 .

Murata M, Cingolani E, McDonald AD, Donahue J, Marban E. Creation of a genetic calcium channel blocker by targeted gem gene transfer in the heart. Circ Res. 2004;95:398-405 .

Nagai K, Murakami T, Iwase T, Tomita T , Sasayama S. Digoxin reduces β-adrenergic contractile response in rabbit hearts. Ca2+-dependent inhibition of adenylylcyclase activity via Na+/Ca2+ exchange. J Clin Invest. 1996; 97: 613-620.

Neumann M., Grieshammer T., Chuvpilo S. RelA/p65 is a molecular target for the immunosuppressive action of protein kinase A. EMBO J. 1995;14:1991-2004.

Olson E, Schneider M. Sizing up the heart development redox in disease. Genes Dev. 2003;17:1937-1956 .

Oudit G, Kassiri Z, Sah R, Ramirez R, Zobel C, Backx P. The molecular physiology of the cardiac transient outward potassium current (I to) in normal and diseased myocardium. J Mol Cell Cardiol. 2001;33:851-872.

O-Uchi J, Sasaki H, Morimoto S, Kusakari Y, Shinji H, Obata T, Hongo K, Komukai K, Kurihara S. Interaction of a1-adrenoreceptor subtypes with different G proteins induces opposite effects on cardiac L-type Ca2+ channel. Circ Res. 2008;102:1378-1388.

Packer M. Current role of beta-adrenergic blockers in the management of chronic heart failure. Am. J. Med. 2001;110(Suppl. 7A):81S–94S.

Palatini P, Casiglia E, Julius S, Pessina A. High heart rate: A risk factor for cardiovascular death in elderly men. Arch Intern Med 1999;159:585-592.

Paterson J, Smith S., Harmar A, Antoni F. Control of a novel adenylyl cyclase by calcineurin. Biochem Biophys Res Commun. 1995;214:1000-1008.

Pedarzani P, Krause M, Haug T, Storm J. Stuhmer W. Modulation of the Ca2+-activated K+ current by a phosphatase–kinase balance under basal conditions in rat CA1 pyramidal neurons. J Neurophysiol. 1998;79:3252-3256.

Peterson B, DeMaria C, Adelman J, Yue D. Calmodulin is the Ca2+ sensor for Ca2+-dependent inactivation of L-type calcium channels. Neuron. 1999;22:549-558.

Petrashevskaya N, Bodi I, Rubio M, Molkentin J, Schwartz A. Cardiac function and electrical remodeling of the calcineurin-overexpressed transgenic mouse. Cardiovasc Res. 2002;54:117-132.

Petrashevskaya N, Ishii M, D'Souza K, Koch S, Fuller-Bicer G, Schwartz A. Presynaptic stimulus-release and postsynaptic compensatory changes in ice lacking the N-type calcium channel á(1b)-subunit. Auton Neurosci. 2011;160:9-15.

Petrashevkaya N, Koch S, Bodi I, Schwartz A. Calcium cycling, historic overview and perspectives. Role for autonomic nervous system regulation. J Mol Cell Cardiol. 2002;34:885-896.

Petrashevskaya N, Gaume B, Mihlbachler K, Dorn G, Liggett S. Bitransgenics with b2-adrenergic receptors or adenylyl cyclase fails to improve β1-adrenergic receptor cardiomyopathy. Clinical Translational Studies 2008 ;1:221-225 .

Pitt G. Calmodulin and CaMKII as molecular switches for cardiac ion channels. Cardiovasc Res . 2007; 73:641-647.

The PKC-DRS Study group. Effect of riboxistaurin in patients with diabetic macular edema: thirty-month results of the randomized PKC-DMES clinical trial. Arch Ophtalmol 2007;125:318-324.

Premont R, Matsuoka I, Mattei M. Identification and characterization of a widely expressed form of adenylyl cyclase. J Biol Chem. 1996;271:13900-13907.

Ravindran A, Qi Zong Lao, Harry JB, Abrahami P, Kobrinsky E, Soladtov N. Calmodulin-dependent gating of Cav1.2?1 calcium channels in the absence of Cav??subunits. Proc. Natl. Acad. Sci. USA. 2008;105: 8154-8159.

Rockman H, Koch W, Lefkowitz R. Seven-transmembrane-spanning receptors and heart function. Nature.2002; 415;206-212.

Richard S, Leclercq F, Lemaire S, Piot C, Nargeot J. Ca2+ currents in compensated hypertrophy and heart failure. Cardiovasc Res. 1998;37:300–311.

Sadoshima J, Morgan J, Izumo S. Angiotensin II, other hypertrophic stimuli mediated by G protein-coupled receptors activate tyrosine kinase, mitogen-activated protein kinase, and 90-kD S6 kinase in cardiac myocytes. The critical role of Ca2+-dependent signaling. Circ Res. 1995;76:1–15.

Sakata K, Shirotani M, Yoshida H, Nawada R, Obayashi K, Togi K, Miho N. Effect of amlodipine and cilnidipine on cardiac sympathetic nervous system and neurohormonal status in essential hypertension. Hypertension. 1999; 33:1447-1452.

Sah R, Ramirewz R, Backx P. Modulation of Ca2+ release in cardic myocytes by changes in repolarization rate:role of phase-1 action potential repolarization in excitation-contraction coupling. Circ Res. 2002;90:165-173.

Serikov V, Bodi I, Koch S, Muth J. Mikala G, Martinov S, Haase H, Scwartz A. Mice with cardiac-specific sequestration of the b-subunit of the L-type calcium channel. Biochem Biophys Res Commun. 202;293:1405-1411.

Shou W, Aghdasi B, Armstrong D. Cardiac defects and altered ryanodine receptor function in mice lacking FKBP12. Nature 1998;391:489-492.

Sudgen P. An overview of endothelin signaling in the cardiac myocytes J Mol Cell Cardiol. 2003; 35:871-886.

Schroder F, Handrock R, Beuckelman D, Hirt S, Hullin R, Priebe L, Schwinger R, Weil J, Herzig S. Increased availability and open probability of single L-type calcium channels from failing compared with nonfailing human ventricle. Circ Res. 1998;98:969-976.

Tedford H, Zamponi G.. Direct G protein modulation of Cav2 channels. Pharmacol Rev 2006;58:837–862.

Takami T, Shigematsu M. Effect of calcium channel antagonists on left ventricular hypertrophy and diastolic function in patients with essential hypertension. Clin Exp Hypertens. 2003;25:523-535.

Takahara H. Koganei, T. Takeda, S. Iwata A. Antisympathetic and hemodynamic property of a dual L/N-type Ca2+ channel blocker cilnidipine in rats. Europ J Pharmacol 2002;434:43-47.

Tsuruta I., Lee H., Masuda N., Koyano-Nakagawa N, Arai N, Arai K, Yokota T. Cyclic AMP inhibits expression of the IL-2 gene through the nuclear factor of activated T cells (NF-AT) site, and transfection of NF-AT cDNA abrogates the sensitivity of EL-4 cells to cyclic AMP. J Immunol. 1998;154:5255-5264.

Valdivia H. Modulation of intracellular Ca2+ levels in the heart by sorcin and FKBP12, two accessory proteins of ryanodine receptors. Trends Pharmacol Sci. 1998;19:479-482.

Wang J, Kelly P. Postsynaptic calcineurin activity downregulates synaptic transmission by weakening intracellular Ca2+ signaling mechanism in hippocampal CA1 neurons. J Neurosci 1997;17:4600-4611.

Wegner M, Cao Z, Rosenfeld M. Calcium-regulated phosphorylation within the leucin zipper of C/EBPβ. Science 1992;256:370-373 .

Wencker D. A mechanistic role for cardiac myocyte apoptosis in heart failure. J. Clin. Invest. 2003;111:1497–1504.

White DC, Hata JA, Shah AS, et al. Preservation of myocardial adrenergic receptor signaling delays the development of heart failure after myocardial infarction. Proc Natl Acad Sci U S A. 2000;97: 5428 –5433.

Wier W, Balke G. Ca2+ release mechanism, Ca2+ sparks, and local control of excitation-contraction coupling in normal heart muscle. Circ Res. 1999;85:770-776.

Winslow R, Rice J, Jafri S, Marban E, O'Rourke B. Mechanisms of altered excitation-contraction coupling in canine tachycardia-induced heart failure: model studies. Circ Res. 1999;84:571-586.

Winstel R, Freund S, Krasel C, Hoppe E, Lohse M. Protein kinase cross-talk: membrane targeting of the β-adrenergic receptor kinase by protein kinase C. Proc Natl Acad Sci USA 1996;93:2105-2109.

Woodcock E, Markovich S. Ins(1,4,5) P3 receptors and inositol Phosphates in the heart-evolutionary artifacts or active signal transducer? Pharmacol Ther. 2005;107:240-251.

Wickenden A, Kaprielian R, Kassiri Z. The role of action potential prolongation and altered intracellular calcium handling in the pathogenesis of heart failure. Cardiovasc Res. 1998;37:312-323.

Wu X, Zhang T, Bossuyt J, McKinsey T, Dedman J, Olson E, Chen J, Brown J, Bers D. Local InsP3-dependent perinuclear Ca2+ signaling in cardiac myocyte exitation-contraction coupling. J Clin Invest. 2006;116:675-682.

Yang L, Chen X, Margulies K, Jeevanandam V, Pollack P, Bayley B, Houser S. L-type calcium channels a1c subunit isoform switching in failing human ventricular myocardium. J Mol Cell Cardiol. 2000;32:973-984 .

Yang L, Katchman A, Morrow J, Doshi D, Marx S. Cardiac L-type calcium channel (CaV1.2) associates with γ subunits. FASEB J. 2011; 928-936.

Yu H., Ma H., Green R. Calcium entry via L-type calcium channels acts as negative regulator of adenylyl cyclase activity and cyclic AMP levels in cardiac myocytes. Mol Pharmacol. 1993;44:689-693.

Yo H., Ma H, Green R. Calcium entry via L-type calcium channels acts as a negative regulator of adenylyl cyclase activity and cyclic AMP levels in cardiac myocytes. Mol Pharmacol. 1993;44:689-693.

Zho Y, Yamazaki T, Nakagawa K, Yamada H, Irigushi N, Toko H. Continuous blockade of L type Ca2+ channels suppresses activation of calcineurin and development of cardiac hypertrophy in spontaneously hypertensive rats. Hypertens Res. 2002;25:117-124.

Zhu W, Wang S, Chakir K, Yang D, Zhang T, Brown J, Devic E, Kobilka B, Cheng H, Xiao RP. Linkage of β1-adrenergic stimulation to apoptotic heart cell death through protein kinase A–independent activation of Ca2+/calmodulin kinase II. J Clin. Invest. 2003;111:617-625.

Zhang C, McKinsey T, Chang S, Antos C, Hill J, Olson E. Class II histone deacetylases act as signal-responsive repressor of cardiac hypertrophy. Cell. 2002;110:479-488.

In: Calcium Channels: Properties, Functions and Regulation   ISBN: 978-1-61470-232-0
Editor: Mark R. Figgins   © 2012 Nova Science Publishers, Inc.

Chapter 3

THE ROLE OF VOLTAGE GATED CALCIUM CHANNELS IN THE DEVELOPING VERTEBRATE RETINA

Elizabeth A. MacMurray and Margaret S. Saha[‡]
Department of Biology
College of William and Mary
Williamsburg, VA, U. S.

Abstract

While the role of voltage-gated calcium channels (VGCCs) in the adult retina is well established, their role in development and in the embryonic retina has received less attention. However, in the developing retina, the diversity of calcium channels, particularly the L-type and T-type, contributes to cell proliferation, differentiation, and synaptogenesis. Additionally, VGCCs may contribute to multiple forms of intracellular calcium activity which regulates developmental processes. L-type channels are functional prior to synapse formation and may regulate optic cup formation. Both T-type and L-type channels are involved in retinal cell proliferation and are possibly involved in the proliferative mechanism known as interkinetic nuclear migration. T-type channels may contribute to retinal progenitor cell differentiation and are typically downregulated during late development. Transient high densities of the T-type channels are associated with undifferentiated retinal progenitors and undifferentiated retinoblastoma cell lines. L-type channels are necessary for retinal synapse formation and establishing retinogeniculate connections via spontaneous retinal waves. The necessity of L-type channels in synaptogenesis is exhibited by mutations in the $Ca_V1.4$ subunit, an essential component for photoreceptor ribbon synapses. Further research on calcium channels may elucidate their role in important aspects of retinal development as well as their potential as therapeutic targets for cell repair.

[‡] E-mail address: mssaha@wm.edu

INTRODUCTION

While the genetics, biochemistry, and physiology of voltage-gated calcium channels have been extensively studied in the adult (reviewed in [1-4]), the role these channels play in the developing organism has received relatively less attention. However, in recent years there has been a growing body of work using a variety of model organisms documenting the importance of voltage-gated calcium channels (VGCCs) during development. Initial research analyzed developmental changes in Ca^{2+} currents by primarily employing electrophysiology techniques, such as the whole-cell patch clamp method. More recently, data garnered from molecular analyses of mRNA and protein expression through *in situ* hybridization and immunocytochemistry has provided a valuable complement to the results obtained via electrophysiological and pharmacological approaches and has enabled researchers to identify specific Ca^{2+} channels responsible for the diversity of Ca^{2+} currents. Each of these techniques permits the identification of Ca^{2+} channel expression and function in multiple organ systems. This review will focus on the role of VGCCs in one of the most well studied organ systems, the developing retina.

Due to the retina's well delineated cell types and the implications for understanding the ontogeny of retinal disease in the adult, there has been significant progress in understanding the molecular and morphological basis of retinal development (reviewed in [5-7]). Early in development, gradients of bone morphogenetic proteins (BMPs) and WNT antagonists along with fibroblast growth factor (FGF) and their downstream targets act to determine the neural and non-neural ectoderm. Combinatorial actions of transcription factors specify an eye-field in the anterior neural plate from which the retinal progenitor cells will emerge. The eye-field is then bisected by the action of sonic hedgehog, leading to the emergence of two optic vesicles. The optic vesicles protrude until they touch the overlying placodal ectoderm and then proceed to form the optic cup. At the optic cup stage, retinal progenitor cells begin to divide extensively in a phase known as proliferation. Following the proliferative phase, the cells then begin to differentiate into mature retinal cells and express proteins that will allow them to carry out their adult retinal function. Simultaneously, synaptogenesis commences and retinal cells begin to elaborate the neural connections that allow visual images to be generated and interpreted by the brain. This chapter will review recent literature that addresses the role of VGCCs at each of the stages of retinal development, namely proliferation, differentiation, and retinogeniculate synaptogenesis. Each section will demonstrate how VGCC RNA, protein, and current expression implicate these channels in both early developmental processes and intracellular calcium activity intrinsic to early developmental stages. This chapter will conclude with future directions for retinal VGCC research.

CA^{2+} CHANNELS IN EARLY DEVELOPMENT: THE PROLIFERATIVE PHASE

While there is little evidence that Ca^{2+} channels are implicated in retinal cell determination, ample studies show VGCCs are involved in cell proliferation. Proliferation is a stage of extensive cell division that begins with the initial mitoses of committed retinal progenitor cells and continues until the last of the neural retinal cell types (i.e. Müller and bipolar cells) are born (reviewed in [5, 8]). During the early proliferative phase, expression of the VGCC α_1 subunits (see Table 1) in vertebrates commences. The α_1 subunit is a protein containing approximately 2,000 amino acids structured into four repeated domains [9]. This subunit creates the channel's pore-forming region and contains the voltage sensor; it is largely responsible for defining the different types of Ca^{2+} current [9, 10]. The mRNA expression of the pore-forming α_1 subunits during early development has been characterized in the developing retina of *Xenopus laevis* embryos using whole-mount *in situ* hybridization [10]. During optic cup formation, *Xenopus* embryos begin RNA expression of the L-type VGCC $Ca_v1.3$ subunit and the P/Q-type $Ca_v2.1$ subunit in the central retina [10]. The R-type $Ca_v2.3$ subunit is not identified in the *Xenopus* retina at any developmental period [10], but has been identified in mammalian retinal progenitor cells [11]. One reason for this discrepancy might be due to alternative splicing of the subunit. RT-PCR and northern blot analyses identify a long variant of the alternatively spliced $Ca_v2.3$ subunit in the inner nuclear layer (INL) of adult Wistar rats [12].

Table 1. Calcium Channel Nomenclature

Channel Number	Subunit Name	Gene	Current Type	HVA or LVA
$Ca_v1.1$	α_{1S}	*CACNA1S*	L	HVA
$Ca_v1.2$	α_{1C}	*CACNA1C*	L	HVA
$Ca_v1.3$	α_{1D}	*CACNA1D*	L	HVA
$Ca_v1.4$	α_{1F}	*CACNA1F*	L	HVA
$Ca_v2.1$	α_{1A}	*CACNA1A*	P/Q	HVA
$Ca_v2.2$	α_{1B}	*CACNA1B*	N	HVA
$Ca_v2.2$	α_{1E}	*CACNA1E*	R	HVA
$Ca_v3.1$	α_{1G}	*CACNA1G*	T	LVA
$Ca_v3.2$	α_{1H}	*CACNA1H*	T	LVA
$Ca_v3.3$	α_{1I}	*CACNA1I*	T	LVA
CACNA2D1	$\alpha_{2\delta}$	*CACNA2D1*	-	-
CACNA2D2	$\alpha_{2\delta}$	*CACNA2D2*	-	-
CACNA2D3	$\alpha_{2\delta}$	*CACNA2D3*	-	-
CACNA2D4	$\alpha_{2\delta}$	*CACNA2D4*	-	-
CACNB1	β_1	*CACNB1*	-	-
CACNB2	β_2	*CACNB2*	-	-
CACNB3	β_3	*CACNB3*	-	-
CACNB4	β_4	*CACNB4*	-	-

Note. HVA = High-voltage activated; LVA = Low-voltage activated

In addition to RNA expression during the proliferative phase of retinal development, research employing electrophysiology and calcium imaging techniques identifies characteristics of functional Ca^{2+} channel currents in proliferating retinal cells. One broad category of current found in retinal progenitor cells is the high-voltage activated (HVA) VGCC current, which at this stage is associated with three subtypes of current: the L-type, P/Q-type, and N-type (see Table 1). The HVA Ca^{2+} channel current has been observed in the retina as early as optic vesicle formation [13, 14], prior to the development of fully differentiated retinal neurons. Early work by Brady and Hilfer [13] reveals the necessity of intracellular calcium for optic cup invagination in embryonic chicks. Their work demonstrates that the inhibition of calcium flux via verapamil, an L-type channel antagonist, prevents optic cup formation. Yamashita and Fukuda [14] followed this work by using Fura-2 calcium imaging to show the presence of intracellular calcium depolarizations during optic cup formation in embryonic day (E) 3 chick retinal slices. The application of nifedipine, another L-type channel antagonist, inhibits these depolarizations, revealing that calcium influx is a product of L-type channel currents. Thus, it appears L-type currents are prominently expressed during early retinal cell proliferation.

Even as some retinal cells begin to differentiate, the proliferation of late-born retinal cell types (see Table 2) continues throughout this period. Autoradiography studies demonstrate that during mammalian retinal development, the early postnatal period is important for the continued proliferation of late cell types, such as Müller glial cells and lens cells [15, 16]. During later stages of proliferation, retinal cells undergo changes in the kinetics of their Ca^{2+} channel currents. The timing of these changes might implicate the low-voltage activated (LVA) T-type Ca^{2+} channels in regulating the proliferative process. Postnatal day (PD) 10 rabbit Müller cells express LVA currents with kinetics similar to those of human T-type channels [17]. Approximately 50% of this inward current can be blocked by the T-type channel antagonist flunarizine. The current is unaffected by the fast Na^+ channel blocker tetrodotoxin (TTX), suggesting that it is primarily a calcium current. The LVA Ca^{2+} channel current amplitudes increase in the earlier postnatal period (PD 7) just as Müller cell proliferation is ending. They then level off to moderately constant levels around PD 14, suggesting that fewer calcium channels are becoming functional during this time period. HVA currents similar to the sustained L-type are also found around this later period, but more prominently after three weeks. Only a small subpopulation of Müller cells expresses both currents simultaneously [17]. The observation that T-type channels are expressed in the time window corresponding with the end of Müller cell proliferation implicates a role for these channels in this process. The late expression of L-type channels also suggests the proliferative role of T-type channels and demonstrates that L-type channels are involved in the later maturation of these cells.

Although not of retinal origin, it is interesting to note that the human lens epithelial cell line (HLE-B3) also displays both VGCC subunit RNA expression and membrane currents and thus may provide additional insights into the role of Ca^{2+} channels in retinal cell proliferation [18]. Proliferation assays reveal two main proliferative stages of HLE-B3 cells: one in which cells exemplify cell growth, passage (P) 9 and one in which cells are static, P16. RT-PCR analyses reveal that in the proliferative state, HLE-3B cells contain T-type subunit mRNA for $Ca_v3.1$, $Ca_v3.2$, and $Ca_v3.3$ (see Table 1) and L-type subunit mRNA for $Ca_v1.2$. Western blot assays demonstrate that T-type channel protein is heavily downregulated in the static cells of P16 as opposed to L-type channel protein which appears unaltered in the static passage [18].

Proliferative HLE-B3 cells also contain both L-type and T-type currents as demonstrated by the whole-cell patch clamp method [18]. The function of these currents in cell proliferation is assessed by utilizing calcium channel antagonists. Blocking the currents with either mibefradil, a T-type and L-type VGCC blocker, or verapamil, an L-type VGCC blocker, results in decreased HLE-B3 proliferation [18]. Additionally, western blot analysis shows that antagonizing the channels results in a downregulation of phosphorylated ERK which activates the cell cycle promoter cyclin D1. These effects may be reversed by use of the L-type channel agonist Bay K 8644. Thus, by possibly leading to the phosphorylation of ERK and activation of cyclin D1, Ca^{2+} channel currents may be an essential component of lens cell proliferation.

Table 2. Developmental Stages of Retinal Cell Birth and Stage of First Cell-specific Ca^{2+} Current Expression

Organism	RGC	Cone	HC	AC	Rod	Muller	Bipolar
Chick	*E2*	*E3* **L-type (EE12-17)***	*E4*	*E4.5* **N-type (EE15-17)***	*E7*	*E9*	*E10*
Mouse	*E11.5* **T-type (E15)** **L-type (E18)** **N-type (E18)**	*E13.5*	*E15.5*	*E17.5*	*P2*	*P4*	*P6*
Rabbit	*E15.5*	*E18.4*	*E21.5*	*E24.5*	*E29.5*	*P1* **T-type (P2)*** **L-type (P24)***	*P4*
Rat	*E13* **T-type (E17)** **L-type (E20)** **N-type (E20)**	*E15.4*	*E17.8*	*E20.3*	*P2.5*	*P5*	*P7*
Xenopus	*St. 22-28*	*St. 22-28*	*St. 22-28*	*St. 28*	*St. 28*	*St. 35-38*	*St. 35-41*
Zebrafish	*28-32 hpf*	*65 hpf*	*50 hpf*	*50 hpf*	*15-18 hpf*	*72-96 hpf*	*60 hpf*

Note. *Represents cell culture. All other Ca^{2+} current findings are from retinal slice preparations. L-type, T-type, N-type refers to approximate first stage of VGCC current expression. Italics delineate stage of cell birthdays. Bold represents stages of currents. RGC = retinal ganglion cells; Cone = cone photoreceptors; HC = horizontal cells; AC = amacrine cells; Rod = rod photoreceptors; Muller = Müller glial cells; EE = embryonic equivalent day; E = embryonic day; P = postnatal day; hpf = hours post fertilization; St.. = stage [6, 15, 17, 30, 31, 70, 71, 79, 80].

As in Müller cells, both L-type and T-type channels may each play a distinct role in the lens cell proliferative mechanism. However, more specific T-type channel antagonists and agonists will be required to determine the exact function of the T-type Ca^{2+} current in this process.

Another potential Ca^{2+} channel-mediated component of proliferation is interkinetic nuclear migration, a Ca^{2+} activity-dependent process in which a mitotic cell's nucleus shifts from the apical surface to the basal surface depending on cell-cycle timing [19, 20]. Studies employing multi-photon imaging demonstrate the necessity for gap junctions in propagating intracellular Ca^{2+} transients, which help coordinate the saltatory movements of retinal progenitor cell nuclei [21]. Although gap junction research does not attempt to establish a role for Ca^{2+} channels in these movements, use of carbenoxolone (CBX), a non-specific gap junction blocker and potential Ca^{2+} channel blocker [21], for these studies could demonstrate a potential role for Ca^{2+} channel influx. Blocking gap junctions with CBX affects coordinated cell movements, and it is possible that Ca^{2+} channels also affect nuclei movements leading to cell proliferation by regulating calcium oscillations. However, recent studies suggest that although T-type channels may be present in proliferating E5 chick retinal ganglion cells (RGCs), these channels (as well as gap junctions) might not be responsible for the synchronous Ca^{2+} oscillations often associated with early developmental processes such as interkinetic nuclear migration [22]. Further research will be needed to elucidate the roles of both gap junctions and VGCCs in retinal cell proliferation.

In conclusion, multiple voltage-gated calcium channels are prominent in both early and late proliferating retinal cells. They are necessary for optic cup formation and additionally regulate the proliferation of Müller glial and lens epithelial cells. VGCCs are clearly involved in cell cycle regulation; whether this applies to interkinetic nuclear migration remains unclear. Future studies are needed to address VGCC protein expression in vertebrates in cell culture and whole retina in combination with Ca^{2+} current kinetic studies.

CA^{2+} CHANNELS IN EARLY DEVELOPMENT: DIFFERENTIATION

Differentiation is the developmental period when cells exit the mitotic cycle and begin expressing the RNA and proteins characteristic of their adult phenotype (reviewed in [5, 23]). In the retina, progenitor cells exiting the cell cycle will become one of six main neural cell types: the retinal ganglion cells (RGCs), horizontal cells, amacrine cells, photoreceptor cells, bipolar cells, and Müller glia (see Table 2). Each of these cells will migrate to their respective layers where they will express unique patterns of VGCCs. These patterns are evident from *in situ* hybridization in *Xenopus* [10]. By the swimming tadpole stages, the *Xenopus* $Ca_v1.3$ RNA expression has moved from the central retina to the photoreceptor layers and is observed in low levels. This pattern is consistent with the *in situ* hybridization patterns of RNA expression observed in mice beginning at E12.5 [10, 24], an embryonic stage of retinal differentiation. In *Xenopus,* the remaining L-type channel subunits, $Ca_v1.1$, $Ca_v1.2$, $Ca_v1.4$, the N-type subunit, $Ca_v2.2$, and the T-type channel subunits, $Ca_v3.1$, $Ca_v3.2$, and $Ca_v3.3$, mRNAs are also expressed at this early differentiation period [10]. Interestingly, in *Xenopus,* the $Ca_v1.4$ subunit mRNA does not appear in the embryo prior to the swimming tadpole stages and is only identified in the retina, specifically in the ganglion cell layer (GCL), the outer nuclear layer (ONL), and within the photoreceptors. Again, these expression patterns are similar to the expression patterns found using immunocytochemistry in four week old chicks [25]. Thus, the location of specific VGCC subunits during differentiation may be partially conserved across species and the locations maintained into adulthood.

Multiple electrophysiology studies are consistent with VGCC subunit RNA expression described during differentiation, particularly in the retinal ganglion cells. Across species, the RGCs are one of the first retinal cell types to differentiate. During this developmental period, L-type, N-type, and transient T-type currents are observed in the developing RGCs of cat and Brown Norway rats [26, 27]. The changes in VGCC current kinetics from the period of cell differentiation to cell maturation might support the role of VGCCs in the differentiation process. Three Ca^{2+} currents, a sustained HVA current, a transient HVA current, and a transient LVA current are present in dissociated primary cultures of embryonic cat RGCs [26]. The sustained HVA current associated with L-type channels has an increased Ca^{2+} channel current density when these cells transition from embryonic cells (E34) to postnatal cells (PD105), a trend observed in other vertebrates [26-28]. A similar finding is also detected for the transient HVA current with the additional observation that the half point of inactivation voltage for these currents significantly decreases between the embryonic period (E39) and the postnatal period. The contribution of this channel to the overall Ca^{2+} current also decreases in prominence in the postnatal period. Schmid and Guenther [27] identify a similar trend in the HVA N-type and P/Q-type currents of rat RGCs from retinal slices. Using the patch clamp technique, they note a transient decrease in duration of Ca^{2+} current deactivation for each Ca^{2+} channel current between the prenatal and postnatal developmental period. It is easy to speculate that the transient decrease in deactivation potential observed for multiple currents could have a function in the transition of these retinal cells to a mature phenotype.

Similar to the HVA currents, the onset timing and expression localization of retinal LVA currents suggest that these currents may regulate cell differentiation. Although the LVA T-type current exhibits no changes in activation or inactivation kinetics from the embryonic to postnatal period in cat RGCs, the current does increase in density during this time [26, 29]. This increase in density could be associated with differentiation or in establishing retinogeniculate connections, as both occur during this developmental period in cats. Research in rat and mouse RGCs would indicate the former [30, 31]. In embryonic rat RGCs, the LVA current is expressed at E17, a developmental period during which retinal cells are not yet fully differentiated. From this stage forward, the LVA current declines and is no longer expressed in adult RGC cells [27, 31]. The decline in LVA current contribution and the increase in HVA current are also observed in mice [30]. Still, it is probable that differences exist in the contribution of T-type channels to retinal differentiation across species, as only a few of the cat RGCs even express this channel [26]. It is also possible that the differences in T-type current are the result of multiple T-type channel subunit expressions. An analysis of these channels at the molecular level may further reveal a reason for the discrepancies amongst mammals.

Other researchers have attempted to identify both a molecular and electrophysiological basis for T-type channel regulation of differentiation. A few remarkable studies utilizing Y-79 retinoblastoma cell lines have demonstrated that during differentiation, retinoblastomas undergo changes in T-type Ca^{2+} current expression, in channel subunit promoters, and in upregulation and downregulation of various T-type channel subunit splice variants [32-35].

Both T-type channel currents and T-type α_1 subunit mRNAs are downregulated after the chemically induced differentiation of these cells [33]. The Ca^{2+} currents in undifferentiated retinoblastomas are identified as T-type due to their transient inward current activation positive to -40 mV, their cessation without external Ca^{2+}, and their ability to be blocked by inorganic cations. Additionally, only the T-type channel blocker mibefradil is successful in reducing the current in comparison to other L-type and N-type channel blockers [33]. Upon differentiation of these retinoblastoma cells on a poly-D-lysine substrate, the inward current is no longer present during depolarizing steps to -40 mV through 0 mV, suggesting that the T-type current has been reduced [33]. This demonstrates that during the differentiation process, both T-type channel mRNA expression and current function are downregulated.

Total RNA extraction from undifferentiated and differentiated retinoblastomas in conjunction with PCR performed on cDNA copies using primers for the human $Ca_v3.1$, $Ca_v3.2$, and $Ca_v3.3$ subunits specifically addresses which T-type channel subunits are present in both cell types [33]. The $Ca_v3.1$ and $Ca_v3.2$ subunits are found in undifferentiated cells and northern blot analyses using the cDNA copies as probes reveals a relative absence of these products in the differentiated cells. The $Ca_v3.3$ subunit is weakly detected in undifferentiated cells and increases slightly in differentiated cells, but not enough to likely contribute to the overall T-type current. The Ca^{2+} current and mRNA detection results support the idea that $Ca_v3.1$ and $Ca_v3.2$ decline during differentiation, suggesting a potential role for them in a cellular differentiation mechanism [33].

Further research demonstrates that transcriptional regulatory elements may drive differences in $Ca_v3.1$ expression between differentiated and undifferentiated retinoblastomas. Cloning a 1.3 kb region upstream from the *CACNA1G* gene (see Table 1) reveals the existence of two potential TATAless promoters (A and B) [34, 35]. Use of 5' RACE shows approximately 500 and 300 bp PCR products in undifferentiated and differentiated cells, respectively. The 500 bp product corresponds to a transcription start site located in promoter A and the 300 bp product has a transcription start site in promoter B [34]. Deletion experiments further demonstrate an enhancer region upstream of promoter A and a repressor region upstream of reporter B, thus depicting a means by which these two promoters may function in differentiated and undifferentiated cells.

T-type channels are potential models for cell differentiation processes in part because they are subjected to alternative splicing in their intracellular regions. Each of their splice variants exhibits different electrophysiological properties, suggesting each one may have different functions in development [2, 36]. Determining how the expression of these variants changes during the differentiation process could elucidate a mechanism by which cell differentiation is occurring. In retinoblastomas, the $Ca_v3.1$ loop links domains III and IV of the protein containing exon 25. This loop undergoes D-type alternative splicing to yield a short exon 25 (variant b) and a long exon 25 (variant a). Exon 26 additionally undergoes alternative splicing to eliminate 54 nucleotides, forming variant c. Differentiated and undifferentiated retinoblastomas might be characterized by their $Ca_v3.1bc$ and $Cav3.1ac$ subunit splice variant expression, respectively [35]. While the $Ca_v3.1ac$ variant appears only in undifferentiated proliferating cells, variant $Ca_v3.1bc$ appears to be the most prominent variant in differentiated retinoblastomas. Thus, a difference in splice variant expression does exist throughout development. How the electrophysiological properties of these variants may contribute to retinal differentiation has yet to be elucidated.

One step towards understanding the function of T-type channels in differentiation is to knock down the channel RNA expression and observe the effect on differentiating cells. Researchers examining undifferentiated retinoblastomas performed knockdowns using $Ca_v3.1$ oligonucleotides specific to a region with no alternative splicing. Although these experiments yield no effect on the Y-79 cell's ability to differentiate without culturing medium nor on the time course of their differentiation in culturing medium, researchers suggest that the multiple splice variants may allow for tissue-specific calcium signaling necessary for mature neuronal phenotypes. Additionally, the lack of effects for the knockdown study could indicate the contribution of other T-type currents, allowing for redundancy in a differentiation mechanism. Further research with double knockdowns of both the $Ca_v3.1$ and $Ca_v3.2$ subunits are necessary to elucidate the contribution each of these subunits makes to cellular differentiation [35].

One final mechanism by which L-type VGCCs may regulate cell differentiation is via patterns of intracellular calcium activity, a well-characterized phenomenon in the developing nervous system [37-42]. In the retina, intracellular calcium activity often occurs in the form of retinal waves, spontaneous bursts of intracellular calcium influx that propagate along a network of cells. Syed et al. [40] found that spontaneous Ca^{2+} waves exist in both the inner retina and the retinal ventricular zone (VZ). While the function of these waves in the inner retina has been related to retinogeniculate formations (reviewed later in this chapter) [43], their presence in the ventricular zone where cells are still dividing indicates they may have a function in cell differentiation [40]. Fura-2 and Fluo-4 AM Ca^{2+} imaging on whole mount retina from E24-E31 rabbits demonstrate that the onset timing of VZ waves is correlated with inner retinal waves, except that the waves are slower. Interestingly, blocking Ca^{2+} channels with Co^{2+} blocks the VZ waves, demonstrating that these waves might be mediated by Ca^{2+} channels. Furthermore, depletion of intracellular Ca^{2+} stores with thapsigargin has no effect on either inner retinal or VZ waves, emphasizing that the source of intracellular Ca^{2+} is most likely through the L-type Ca^{2+} channels. One implication of these findings is that inner retinal waves may be signaling to the VZ cells to induce differentiation. This raises the possibility they may be mediating a kind of retrograde differentiation process.

However, the role of Ca^{2+} channels in developing VZ cells may be disputed by earlier research examining intracellular Ca^{2+} transients in the VZ of E4/E6 chick retina [41]. Using Fluo-4 AM imaging on whole retinas, researchers find that Ca^{2+} transients induced by muscarinic and purinergic stimulation are unaffected by the Ca^{2+} channel blocker Ni^{2+} and are further suppressed by depletion of their intracellular stores with caffeine. However, GABA- and glutamate-induced transients are affected by Ca^{2+} channel blockers and depletion of intracellular stores has no effect. Further experimentation examining Hoechst staining and neuronal tubulin antibody staining in imaged cells reveals that only cells with muscarinic receptors ubiquitously present are differentiating and proliferating [41]. This indicates that although Ca^{2+} channels are involved in transient calcium activity, they are not necessarily associated with the activity responsible for cellular differentiation.

Even amidst these conflicting results, VGCCs are clearly expressed and functional during developmental differentiation. The onset timing, current kinetic changes, and splice variant possibilities of T-type and L-type channels indicate these channels may contribute to a cellular differentiation mechanism, though studies in retinoblastomas demonstrate research has yet to identify an actual function of VGCCs in this process. Future research will need to knock down multiple calcium channels in retinal specific tissues to identify if there is an effect of these channels on the differentiation process.

CA^{2+} CHANNELS IN REGENERATION AND DEGENERATION

Another approach to understanding the differentiation process is to examine VGCC currents in both dedifferentiating cells and regenerating cells. In both paradigms, cells are undergoing either reverse developmental processes or processes similar to those of early development (reviewed in [44]). Recent studies have demonstrated that L-type Ca^{2+} channels are involved in pathways of dedifferentiation in the chicken retinal pigment epithelium (RPE) [45]. Chick RPE cells dedifferentiate into intermediate cells that may either revert back to RPE or form differentiated lens and retinal neural cells. In examining the muscarinic and nicotinic acetylcholine responses in these dedifferentiated intermediates, researchers have identified a pathway by which nicotinic acetylcholine-induced intracellular calcium influx is allowed via L-type Ca^{2+} channels. This is in contrast to the muscarinic response which releases calcium from intracellular stores. Researchers tested the differences in muscarinic and nicotinic acetylcholine-induced Ca^{2+} influx between differentiated RPE cells and dedifferentiated intermediates grown on EPF medium. Surprisingly, Fluo-4 AM Ca^{2+} imaging shows the muscarinic Ca^{2+} response is present in both differentiated and dedifferentiated RPE cells, but that the nicotinic-dependent Ca^{2+} response is only observed in the dedifferentiated intermediates [45]. Cells that have dedifferentiated are effectively in a state similar to that of undifferentiated cells during development. Since the nicotinic response is mediated by L-type Ca^{2+} channels and is only seen in dedifferentiated intermediates, these findings implicate the channels as important components in the cellular differentiation process.

Cell regeneration and degeneration studies also implicate L-type channels in developmental processes, but with slightly conflicting results. While L-type channels are necessary for cell regeneration and growth, they are also responsible for cell degeneration and must be blocked to reduce either necrotic or apoptotic processes. The adult newt is an exemplary model to demonstrate the role of L-type Ca^{2+} channels in regenerating retinal ganglion cells [46]. Fura-2 imaging reveals that GABA-induced increases in intracellular Ca^{2+} of regenerating, mitotic RGCs is blocked by the L-type Ca^{2+} channel blocker verapamil and by removal of extracellular calcium [46]. Additionally, this GABA-induced depolarizing rise in intracellular calcium is only observed in these mitotic regenerating cells and is not seen in the fully regenerated RGCs. This suggests that the use of these L-type channels is not merely specific to regenerating cells, but rather to cells that have not yet fully differentiated. Because neurotransmitters such as GABA have been implicated in cell proliferation and adult neurotransmitter differentiation [15, 47], it is possible that the GABA depolarization observed in regenerating cells is occurring during a process akin to cellular differentiation. Thus, it is likely that L-type channels are actually contributing to the differentiation process.

In contrast to regeneration, retinal degeneration studies provide an alternative insight into the function of VGCCs in retinal development [48-52]. Research examining retinal neuroprotection demonstrates that secondary neuronal cell death may be related to L-type Ca^{2+} channel expression. The L-type Ca^{2+} channel blocker Lomerizine exhibits neuroprotective effects against neuronal degeneration in both RGCs and visual centers such as the superior colliculus (SC) and the dorsal lateral geniculate nucleus (dLGN) [48, 49]. In Piebald-Virol-Glaxo rats, secondary RGC death induced by damage to the optic nerve is partially rescued by Lomerizine, as assessed through anterograde and retrograde labeling [48]. Lomerizine primarily reduces necrotic RGC death, but also decreases apoptosis in these cells. If inhibition of L-type VGCCs results in retina neuroprotection, it is likely that L-type channels contribute to degenerative processes. However, in more recent studies [50], the extent of Lomerizine's protective effects in RGC damage has been questioned due to the drug's inability to completely restore visual function after either one week or one month of treatment. Additionally, the drug was unable to maintain restoration of the fast reset phase of optokinetic nystagmus. Similar to the Lomerizine findings, the L-type channel blocker nifedipine fails to significantly reduce neuronal damage in a model of retinal ischemia [52]. Despite these dual findings in the RGCs, Lomerizine has been verified to partially prevent secondary SC and dLGN neuronal degeneration after induced retinal damage via intravitreal injection of NMDA in early postnatal mice [49]. Thus, L-type Ca^{2+} channels may have a function in trans-synaptic neuronal protection, but might not be responsible for neural retinal degeneration. These findings would be more consistent with the function of L-type channels in retinal regeneration.

Despite the contradictory findings in RGCs, L-type VGCCs may play a prominent role in photoreceptor degeneration and may furthermore regulate developmentally important processes such as cell apoptosis [51, 53]. This has been observed in *rd1* mice whose rod cells contain a phosphodiesterase-6 beta subunit mutation (*rd1* mice). These mice experience abnormal increases of intracellular calcium leading to degenerating apoptotic cell death in early postnatal development [51]. The use of dihydropyridine L-type channel antagonists is successful in delaying *rd1* cell degeneration, but not in removing it. Microarray analyses comparing mutant *rd1* rod cells treated with L-type VGCC blocker nilvadipine to control *rd1* cells show that treated *rd1* cells exhibit a downregulation in genes related to the apoptotic pathway such as caspases-3,-9, and -14 [53]. These results indicate that L-type channels may function in gene transcription pathways responsible for degeneration and apoptosis. Although the research appears to present two opposing roles for L-type channels in regeneration and degeneration, these roles may be equally important in developmental processes. While the presence of L-type channels in the regenerating retina suggests these channels may contribute to differentiation, their presence in degenerating neurons may be indicative of their regulation of apoptosis, a developmentally important mechanism for axon pruning [54] during synaptogenesis.

CA^{2+} CHANNELS IN SYNAPTOGENESIS

Synaptogenesis is the developmental stage characterized by synapse formation within the neural retina. It is also a period during which retinogeniculate and retinocollicular connections mature (reviewed in [55, 56]). These connections will be essential for neurotransmitter communication in the adult retina. The expression patterns of voltage-gated calcium channels throughout the retina and the brain's visual centers suggest they contribute to this stage of development and are necessary for proper formation of retinal plexiform layers [57]. As previously described, *in situ* hybridization shows the VGCC α_1 subunits are already expressed by synaptogenesis [10], suggesting these channels could additionally contribute to synapse formation. Similar research in zebrafish demonstrates that RNA transcripts for the β subunits are also present during this time [58]. Furthermore, the increasing expression of both α_1 and β subunits in visual centers such as the dorsal lateral geniculate nucleus (dLGN) and the superior colliculus (SC) prior to synapse maturation suggests a functional role for VGCCs during this period.

The cytoplasmic VGCC β subunits are responsible for establishing channel kinetics and voltage-dependence gating [9, 57, 58]. In zebrafish, the β_2, β_3, and β_4 subunit genes (see Table1) are duplicated with each homolog having a different expression pattern throughout the multiple retinal layers [58]. The β_{2a} subunit is expressed 96 hours post fertilization (hpf) in the photoreceptor layer and outer half of the inner nuclear layer (INL). In contrast, its homolog β_{2b} is strongly expressed earlier at 48 hpf in the ganglion cell layer (GCL) and more weakly in the INL. The β_{3a} and β_{3b} subunits are expressed at 72 and 96 hpf in the GCL with some expression of the first homolog in the INL. The β_{4a} subunit is found in the innermost tier of the INL and the β_{4b} subunit is found in both the INL and GCL [58]. The range of expression for the β subunits emphasizes their importance for calcium channel function and also suggests a mechanism of redundancy which may allow VGCCs to maintain cell functioning even if one subunit becomes nonfunctional.

In contrast to the expression patterns of β_3 and β_4 subunits, the β_2 subunits are expressed in regions which do not overlap with the other β subunits. This could suggest that mutations in β_2 subunits will result in a more severe phenotype due to lack of channel redundancy [58]. This has been demonstrated in studies of β_2 null mice [57, 59]. Immunocytochemistry staining and histology reveal these mice suffer from an abnormal distribution in their retinal $Ca_V1.4$ subunit protein patterning and a malformation in their retinal outer plexiform layer. This emphasizes a necessary relationship between VGCC subunits for proper retinal organization, particularly in synapse formation. These findings also implicate at least one L-type channel in these processes.

This is further emphasized by the fact that multiple developmental gene mutations in retinal L-type Ca^{2+} channels and Ca^{2+} channel binding proteins are linked to malformations in the connections between secondary neurons of the photoreceptor layer and the outer nuclear layer [60-65]. Many of these deformities result from mutations in the gene *CACNA1F* which codes for the $Ca_V1.4$ subunit and is associated with congenital stationary night blindness type II (CSNB-2). This disease is characterized by myopia, nystagmus, and abnormal electroretinograms in humans and mammalian models [25, 65]. One excellent model for this disease is the *nob2* mouse whose genome contains a disruption by a transposable element in exon 2 of *CACNA1F*. This mutation results in symptoms characteristic of CSNB-2 in humans [61, 66]. In wildtype mice, connections between secondary retinal neurons occur in the outer plexiform layer (OPL), the layer reserved for cell ganglia. In the *nob2* model, however, ectopic synaptic contacts occur between rods, rod bipolar cells, and horizontal cells in the outer nuclear layer (ONL), the layer primarily reserved for cell bodies [61]. Interestingly, the ectopic connections in the ONL still maintain much of their synaptic machinery, suggesting that VGCCs may not be necessary for the formation of synaptic specializations. Furthermore, in *nob2* mice, the OPL develops normally postnatally and the ectopic synaptic expression in the ONL does not occur until after eye-opening, leaving questions as to the function of L-type channels prior to the formation of ectopic connections. Although the synapse specializations form in the absence of L-type channels, it must be noted that ribbon synapses are malformed

[61]. Thus, it appears the symptoms related to the *CACNA1F* mutation primarily result from the abnormally localized synaptic connections with impaired synapses.

Further research examining the contribution of L-type channels to retinal synaptogenesis has demonstrated L-type subunit protein expression in the brain's visual centers. To establish if Ca^{2+} channels develop in the murine brain during the time of retinocollicular and retinogeniculate formation (PD4-PD15), Mize et al. [67] utilized $Ca_v1.2$ antibodies to stain these L-type VGCC subunits in sections of the lateral geniculate nucleus (LGN), superior colliculus (SC), visual cortex, hippocampus, and cerebellum from PD2-PD28 mice. In all regions, $Ca_v1.2$ protein is labeled by PD4, suggesting the onset of L-type channel expression does occur during the period of retinogeniculate formation. Most interesting is the change in expression of $Ca_v1.2$ in the SC. In this region, $Ca_v1.2$ protein begins as a band in the superficial gray layer. After PD15, the period when retinal connections are fully maturing [68], the expression of this band decreases and the VGCCs become more uniformly dispersed throughout the SC [67]. This expression pattern emphasizes that the region-specific change in $Ca_v1.2$ is concurrent with the maturation of retinocollicular and retinogeniculate projections [67]. Thus, it is likely that L-type Ca^{2+} channels contribute to the organization of these connections.

Although the RNA and protein expression of VGCCs is present during the period of synapse formation, the regulation of this process by VGCCs could not be assessed without examining current kinetics at these stages. The function of VGCCs is demonstrated by electrophysiology performed on retinal reaggregation cultures, amacrine cell cultures, and photoreceptor cell cultures. The findings of these studies support the role of L-type currents in synapse formation and also reveal a role for P/Q-type and N-type currents (see Table 1) [28, 69-71]. In each of these studies, the HVA currents are separated by their pharmacological properties. L-type channels exhibit dihydropyridine sensitivity while P/Q-type and N-type currents are identified by their ω-agatoxin and ω-conotoxin sensitivity, respectfully [2, 27]. By separating these currents, early research was able to identify and analyze each current's expression during the period of retinal synaptogenesis. Furthermore, these studies allowed more recent experiments to demonstrate the role of VGCCs in intracellular calcium waves necessary for establishing retinotopic maps.

Retinal reaggregation cultures (spheroid cultures) have been a popular tool used to analyze HVA current dynamics in developing embryonic retinal cells. Retinospheroid cultures isolated from E6 and E8 chicks display unique L-type and N-type Ca^{2+} channel currents during the *in vivo* equivalent stage of synaptogenesis [28]. Using Indo-1/AM calcium dye, researchers have demonstrated the presence of functional Ca^{2+} channels in E8 chick retinospheroids cultured for 14 days. At day one *in vitro*, retinal cells display very little intracellular calcium activity, suggesting that functional currents are not yet established. The calcium activity gradually increases during development and peaks at the stage akin to synaptogenesis. Verifying that the dye-bound calcium is a result of VGCCs, the intracellular Ca^{2+} is blocked by the L-type channel antagonist nitrendipine. Application of this antagonist reduces Ca^{2+} activity, supporting the role of L-type channels during synapse formation.

Further analysis of retinospheroid L-type VGCCs using [³H] nitrendipine binding assays demonstrates that while the binding affinity of nitrendipine to Ca^{2+} channels does not change with retinospheroid days *in vitro*, the maximum number of binding sites for nitrendipine increases during this time [28]. This indicates that either the number of L-type channels on the plasma membrane is increasing or more channels already present on the membrane are becoming functional. In addition to the rise in L-type channels, the number of N-type channels expressed also increases, as demonstrated using the N-type specific antagonist, ω-conotoxin GVIA. Again, the increase in N-type current correlates with the hatching stages of chick development, a period known for synapse formation. Taken together, these results suggest that both L-type and N-type Ca^{2+} channels may be necessary for establishing the early neural retina synapses necessary for the transmission of light into electric signal. The role these currents may play in retinogeniculate connections has yet to be elucidated.

Similar to chick retinospheroids, primary cultures of GABAergic amacrine cells isolated from E8 chicks develop VGCC current expression at the developmental stage when cells are reaching maturity. Additionally, when subjected to patch clamp analysis, these cells demonstrate an HVA current pattern resembling L-type VGCC currents [69]. This current exhibits inward expression of a sustained nature which is blocked by the addition of Co^{2+}. After being cultured *in vitro* for 6-9 days, the putative amacrine cells display similar current recordings to those of *in vivo* E17 retinal tissue slices. Prior to six days in culture, the cells exhibit no voltage-gated currents. Thus, these cells are experiencing changes during development similar to their whole retina counterparts, demonstrating that it is only as these cells develop mature retinal connections that they express L-type VGCCs. It is interesting to note that while these GABAergic cultured cells (identified by their [³H] muscimol uptake and GABA immunoreactivity) exhibit L-type currents, recent experiments have suggested that in adult mice, cholinergic amacrine cells express HVA P/Q-type and N-type currents [72]. These differences in expression could indicate unique roles for these currents in individual retinal cell subtypes. Future research should explore the role P/Q-type and N-type currents in developing cholinergic amacrine cells to establish if they are present during a similar developmental period.

Avian photoreceptors also display multiple HVA current kinetics which may relate to the development of specific cone cell subtypes prior to or during synaptogenesis [70, 71]. After three days in culture, E8 chick cone cells begin to display L-type Ca^{2+} currents and develop prominent inward Ca^{2+} current expression by embryonic equivalent (EE) day 16 [70]. Whole-cell patch clamp analyses demonstrate the inward Ca^{2+} currents are blocked by Co^{2+} and are enhanced when extracellular Ba^{2+} is used as the charge carrier in replace of calcium. Additionally, the Ca^{2+} channel currents are enhanced with the agonist Bay K 8644 and knocked down by L-type channel antagonist nifedipine. Intracellular calcium recordings using E6 chick photoreceptors cultured for 6-12 days display similar L-type properties both in response to nifedipine and Bay K 8644 [71]. These findings demonstrate that by the embryonic equivalent period of inner retinal synaptogenesis, cone cells are expressing prominent L-type VGCC expression.

As expected, the Ca^{2+} current density in cone photoreceptors does not remain constant during development, but rather increases as more Ca^{2+} channels are likely added to the plasma membrane during cell maturation [70]. Fura-2 imaging further emphasizes the time-scale of cone photoreceptor VGCC development with cells exhibiting increases in intracellular cytosolic calcium upon depolarization with extracellular 100 mM K^+ as they develop from EE10-EE16. Specifically, in comparing the intracellular calcium responses of EE10 cells to those of EE14, only EE14 display activation of inward Ca^{2+} current. Interestingly, researchers have found a range of calcium current activation voltages amongst multiple cone cells [70]. The cells activate at voltages between -70 mV and -25 mV, with either slow inactivation or no inactivation present. This suggests that individual currents may be specific to individual photoreceptor cell types and may also represent a mechanism by which photoreceptors are distinguished from other retinal cell types.

Although earlier research has established the presence of calcium channel currents in maturing cells, it is only in more recent studies that researchers have demonstrated a function of these channels. Specifically, L-type Ca^{2+} channels may be influencing retinogeniculate connections via retinal waves or calcium activated gene transcription [43, 73, 74]. Retinal waves drive increases in cyclic nucleotides required for growth cone repulsion and thus may contain information in their burst pattern that leads to appropriate ephrin concentration gradients. These gradients are necessary to help establish retinotopic maps [74]. L-type Ca^{2+} channels may play a particularly important role in retinal waves due to the findings that L-type channel agonists have non-synaptic spatial and temporal effects on these waves in the RGCs of postnatal mice [43]. Researchers have identified retinal waves in which L-type channel agonists (Bay K 8644) could prevent gap junction blockers from reducing the velocity with which these waves travel. Interestingly, these L-type Ca^{2+} channel-mediated waves are not affected by TTX, indicating that they travel in the absence of fast synaptic transmission. Thus, these L-type Ca^{2+} channel-mediated waves might be a mechanism by which diverse neurotransmitter systems can propagate the spontaneous activity necessary to form or refine retinotopic connections [43].

An alternative mechanism by which the L-type channel-mediated calcium activity could be affecting axon pathfinding and synapse formation is through CRE/CREB-mediated transcription activation, which is also known to increase during the early mammalian postnatal period [73, 75]. L-type Ca^{2+} channels convey calcium signals to the nucleus by binding calmodulin at the carboxyl terminus of the Ca^{2+} channel protein [75]. This activates the MAPK pathway which leads to the transcription of genes involved in synaptogenesis. The presence of CRE/CREB-mediated transcription during the establishment of retinogeniculate connections could suggest it contributes to the regulation of this process. However, recent research using E18 mice lacking CREB in neural progenitors suggest that a downregulation of the CREB pathway actually contributes to axon growth [76]. This indicates that L-type channels may possibly regulate non-synaptic connections. Whether or not this function of L-type channels is related to retinal cells will have to be revealed by future research.

In conclusion, VGCCs are prominently expressed during vertebrate synapse formation. These patterns of expression indicate that both L-type and N-type channels contribute to establishing retinogeniculate synapses, as well as neural retinal connections. The role of L-type VGCCs is further emphasized by the knockdown of α_1 and β subunits in retinal mutations. Furthermore, the contribution of L-type channels to calcium activity could implicate them in gene transcription pathways leading to axon pathway refinement.

CONCLUSION

While recent studies have demonstrated that VGCCs play an important role during embryonic development, many questions remain. For example, further research is essential for establishing if Ca^{2+} channels contribute to stages of retinal development prior to proliferation, such as retinal cell determination. Careful analyses using sensitive molecular techniques such as RT-PCR should reveal if VGCCs are expressed at the earliest stages of development. It will also be important to conduct comparative studies on multiple model systems at equivalent stages to demonstrate the potential conservation of these developmental mechanisms. Studies which combine the molecular and electrophysiological approaches on the same cells will be particularly powerful, given that these will correlate specific gene expression with the resulting functional currents.

Ca^{2+} channel regulation in the developing embryo is another important future focus. Few experiments have analyzed upstream and downstream regulators of Ca^{2+} channel genes which could be responsible for the changing expression of these channels during the developmental period. However, some studies have focused on the role microRNAs may play in regulating the daily oscillatory patterns of $Ca_v1.2$ mRNA in chick cone photoreceptors [77]. The somatostatin receptor subtype 4 may also regulate Ca^{2+}as its activation has been found to inhibit L-type channels in PD7 rat retinal ganglion cells [78]. Whether these two factors or other factors contribute to regulation in early development remains unknown.

The conflicting literature over the proposed mechanisms through which Ca^{2+} channels contribute to cell proliferation and retinogeniculate synapse formation suggests that further research is required to understand the roles VGCCs play in these processes as well. The development of additional Ca^{2+} channel-specific pharmacological agents will allow the identification of each channel during specific developmental periods. Tissue-specific knockdowns similar to those used in retinoblastoma research could further elucidate some of the VGCC contributions to proliferation and differentiation in specific retinal cell types such as RGCs and amacrine cells. Conditional knockout studies using cre-mediated recombinase technology would also be useful in understanding the tissue-specific function of Ca^{2+} channels. Furthermore, these knockout studies could reveal additional roles for the channels in the spontaneous calcium activity which regulates developmental processes such as interkinetic nuclear migration.

It is clear that without VGCCs, their subunit variants, and their associated proteins, the retina and its visual pathways could not form and function properly. Thus, Ca^{2+} channels are excellent targets for cell repair and regeneration studies. By further examining both the current kinetics and the molecular properties of these channels, researchers will potentially unravel new therapies applicable to retinal disease. Additionally, understanding the interplay between multiple Ca^{2+} channels in each retinal cell type may assist in identifying novel treatment targets for developmental retinal malformations. Thus, the field of VGCC activity in retinal development remains ripe for future progress.

ACKNOWLEDGMENTS

We thank Brittany Lewis for her suggestions on the manuscript. Support was provided by a grant from the Howard Hughes Medical Institute Undergraduate Science Education Program to the College of William and Mary and from the NIH (R15NS067566).

REFERENCES

1. Dolphin, A.C. (2009). Calcium channel diversity: multiple roles of calcium channel subunits. *Current Opinion in Neurobiology, 19*, 237-244.

2. Huc, S. et al. (2009). Regulation of T-type calcium channels: signaling pathways and functional implications. *Biochimica et Biophysica Acta (BBA) - Molecular Cell Research, 1793*, 947-52.

3. Doan, L. (2010). Voltage-gated calcium channels and pain. *Techniques in Regional Anesthesia and Pain Management, 14*, 42-47.

4. Barbado, M., Fablet, K., Ronjat, M. & De Waard, M. (2009). Gene regulation by voltage-dependent calcium channels. *Biochimica et Biophysica Acta (BBA) - Molecular Cell Research, 1793*, 1096-1104.

5. Reese, B.E. (2010). Development of the retina and optic pathway. *Vision Research, In Press, Corrected Proof.*

6. Sernagor, E., Eglen, S., Harris, B., & Wong, R. (ed.) (2006). *Retinal Development.* Cambridge: Cambridge University Press.

7. Zaghloul, N.A., Yan, B. & Moody, S.A. (2005). Step-wise specification of retinal stem cells during normal embryogenesis. *Biology of the Cell, 97*, 321-337.

8. Dyer, M.A. & Cepko, C.L. (2001). Regulating proliferation during retinal development. *Nature Reviews Neuroscience, 2*, 333-42.

9. Catterall, W.A. (2000). Structure and regulation of voltage-gated Ca2+ channels. *Annual Review of Cell and Developmental Biology, 16*, 521-55.

10. Lewis, B.B. et al. (2009). Cloning and characterization of voltage-gated calcium channel alpha1 subunits in Xenopus laevis during development. *Developmental Dynamics, 238*, 2891-2902.

11. Das, A.V. et al. (2005). Membrane properties of retinal stem cells/progenitors. *Progress in Retinal and Eye Research, 24*, 663-681.

12. Kamphuis, W. & Hendriksen, H. (1998). Expression patterns of voltage-dependent calcium channel alpha 1 subunits (alpha 1A-alpha 1E) mRNA in rat retina. *Brain Research. Molecular Brain Research, 55*, 209-220.

13. Brady, R.C. & Hilfer, S.R. (1982). Optic cup formation: a calcium-regulated process. *Proceedings of the National Academy of Sciences of the United States of America, 79*, 5587-5591.

14. Yamashita, M. & Fukuda, Y. (1993). Calcium channels and GABA receptors in the early embryonic chick retina. *Journal of Neurobiology, 24*, 1600-1614.

15. Martins, R.A. & Pearson, R.A. (2008). Control of cell proliferation by neurotransmitters in the developing vertebrate retina. *Brain Research, 1192*, 37-60.

16. Young, R.W. (1985). Cell proliferation during postnatal development of the retina in the mouse. *Developmental Brain Research, 21*, 229-239.

17. Bringmann, A., Schopf, S. & Reichenbach, A. (2000). Developmental regulation of calcium channel-mediated currents in retinal glial (Muller) cells. *Journal of Neurophysiology, 84*, 2975-2983.

18. Meissner, A. & Noack, T. (2008). Proliferation of human lens epithelial cells (HLE-B3) is inhibited by blocking of voltage-gated calcium channels. *Pflügers Archiv, 457,* 47-59.

19. Frade, J.M. (2002). Interkinetic nuclear movement in the vertebrate neuroepithelium: encounters with an old acquaintance. *Progress in Brain Research, 136,* 67-71.

20. Baye, L.M. & Link, B.A. (2008). Nuclear migration during retinal development. *Brain Research, 1192,* 29-36.

21. Becker, D.L. et al. (2007). Multiphoton imaging of chick retinal development in relation to gap junctional communication. *The Journal of Physiology, 585,* 711-719.

22. Yamashita, M. (2008). Synchronous Ca2+ oscillation emerges from voltage fluctuations of Ca2+ stores. *FEBS Journal, 275,* 4022-4032.

23. Davis, D.M. & Dyer, M.A. (2010). In Invertebrate and Vertebrate Eye Development. *Elsevier Academic Press Inc,* 175-188.

24. Kersten, F.F. et al. (2010). Association of whirlin with Cav1.3 (alpha1D) channels in photoreceptors, defining a novel member of the usher protein network. *Investigative Ophthalmology & Visual Science, 51,* 2338-46.

25. Firth, S.I., Morgan, I.G., Boelen, M.K. & Morgans, C.W. (2001). Localization of voltage-sensitive L-type calcium channels in the chicken retina. *Clinical & Experimental Ophthalmology, 29,* 183-187.

26. Huang, S.J. & Robinson, D.W. (1998). Activation and inactivation properties of voltage-gated calcium currents in developing cat retinal ganglion cells. *Neuroscience, 85,* 239-247.

27. Schmid, S. & Guenther, E. (1999). Voltage-activated calcium currents in rat retinal ganglion cells in situ: changes during prenatal and postnatal development. *The Journal of Neuroscience, 19,* 3486-3494.

28. Capela, A., Cristovao, A., Carvalho, C. & Carvalho, A.P. (1997). Ontogeny of the L-type voltage sensitive calcium channels in chick embryo retinospheroids. *Brain Research. Molecular Brain Research, 104,* 63-69.

29. Robinson, D.W. & Wang, G.-Y. (1998). Development of intrinsic membrane properties in mammalian retinal ganglion cells. *Seminars in Cell & Developmental Biology, 9,* 301-310.

30. Rorig, B. & Grantyn, R. (1994). Ligand- and voltage-gated ion channels are expressed by embryonic mouse retinal neurones. *Neuroreport, 5,* 1197-1200.

31. Schmid, S. & Guenther, E. (1996). Developmental regulation of voltage-activated Na+ and Ca2+ currents in rat retinal ganglion cells. *Neuroreport, 7,* 677-681.

32. Barnes, S. & Haynes, L.W. (1992). Low-voltage-activated calcium channels in human retinoblastoma cells. *Brain Research, 598,* 19-22.

33. Hirooka, K. et al. (2002). T-Type calcium channel alpha1G and alpha1H subunits in human retinoblastoma cells and their loss after differentiation. *Journal of Neurophysiology, 88,* 196-205.

34. Bertolesi, G.E. et al. (2003). Regulation of α1G T-type calcium channel gene (CACNA1G) expression during neuronal differentiation. *European Journal of Neuroscience, 17*, 1802-1810.

35. Bertolesi, G.E. et al. (2006). Ca(v)3.1 splice variant expression during neuronal differentiation of Y-79 retinoblastoma cells. *Neuroscience, 141*, 259-268.

36. Chemin, J., Monteil, A., Bourinet, E., Nargeot, J. & Lory, P. (2001). Alternatively spliced alpha(1G) (Ca(V)3.1) intracellular loops promote specific T-type Ca(2+) channel gating properties. *Biophysical Journal, 80*, 1238-1250.

37. Spitzer, N.C. & Gu, X. (1997). Purposeful patterns of spontaneous calcium transients in embryonic spinal neurons. *Seminars in Cell and Developmental Biology, 8*, 13-19.

38. Spitzer, N.C., Root, C.M. & Borodinsky, L.N. (2004). Orchestrating neuronal differentiation: patterns of Ca2+ spikes specify transmitter choice. *Trends in Neurosciences, 27*, 415-421.

39. Moody, W.J. & Bosma, M.M. (2005). Ion Channel Development, Spontaneous Activity, and Activity-Dependent Development in Nerve and Muscle Cells. *Physiological Reviews, 85*, 883-941.

40. Syed, M.M., Lee, S., He, S. & Zhou, Z.J. (2004). Spontaneous Waves in the Ventricular Zone of Developing Mammalian Retina. *Journal of Neurophysiology, 91*, 1999-2009.

41. Pearson, R., Catsicas, M., Becker, D. & Mobbs, P. (2002). Purinergic and Muscarinic Modulation of the Cell Cycle and Calcium Signaling in the Chick Retinal Ventricular Zone. *The Journal of Neuroscience, 22*, 7569-7579.

42. Stacy, R.C., Demas, J., Burgess, R.W., Sanes, J.R. & Wong, R.O.L. (2005). Disruption and Recovery of Patterned Retinal Activity in the Absence of Acetylcholine. *The Journal of Neuroscience, 25*, 9347-9357.

43. Singer, J.H., Mirotznik, R.R. & Feller, M.B. (2001). Potentiation of L-type calcium channels reveals nonsynaptic mechanisms that correlate spontaneous activity in the developing mammalian retina. *The Journal of Neuroscience, 21*, 8514-8522.

44. Lamba, D.A., Karl, M.O. & Reh, T.A.(2009) *Neurotherapy: Progress in Restorative Neuroscience and Neurology*. Amsterdam: Elsevier Science Bv.

45. Sekiguchi-Tonosaki, M., Obata, M., Haruki, A., Himi, T. & Kosaka, J. (2009). Acetylcholine induces Ca2+ signaling in chicken retinal pigmented epithelial cells during dedifferentiation. *American Journal of Physiology - Cell Physiology, 296*, C1195-1206.

46. Ohmasa, M. & Saito, T. (2004). GABAA-receptor-mediated increase in intracellular Ca2+ concentration in the regenerating retina of adult newt. *Neuroscience Research, 49*, 219-227.

47. Tozuka, Y., Fukuda, S., Namba, T., Seki, T. & Hisatsune, T. (2005). GABAergic Excitation Promotes Neuronal Differentiation in Adult Hippocampal Progenitor Cells. *Neuron, 47*, 803-815.

48. Fitzgerald, M. et al. (2009). Secondary retinal ganglion cell death and the neuroprotective effects of the calcium channel blocker lomerizine. *Investigative Ophthalmology & Visual Science 50*, 5456-5462.

49. Ito, Y. et al. (2010). Lomerizine, a Ca2+ channel blocker, protects against neuronal degeneration within the visual center of the brain after retinal damage in mice. *CNS Neuroscience & Therapeutics, 16*, 103-114.

50. Selt, M., Bartlett, C.A., Harvey, A.R., Dunlop, S.A. & Fitzgerald, M. (2010). Limited restoration of visual function after partial optic nerve injury; a time course study using the calcium channel blocker lomerizine. *Brain Research Bulletin, 81*, 467-471.

51. Barabas, P., Cutler Peck, C. & Krizaj, D. (2010). Do Calcium Channel Blockers Rescue Dying Photoreceptors in the Pde6b (rd1) Mouse? *Advances in Experimental Medicine and Biology, 664*, 491-499.

52. Massote, P.D. et al. (2008). Protective effect of retinal ischemia by blockers of voltage-dependent calcium channels and intracellular calcium stores. *Cellular and Molecular Neurobiology, 28*, 847-856.

53. Takano, Y. et al. (2004). Study of drug effects of calcium channel blockers on retinal degeneration of rd mouse. *Investigative Ophthalmology & Visual Science 45*, 5084.

54. Vanderhaeghen, P. & Cheng, H.J. (2010). Guidance molecules in axon pruning and cell death. *Cold Spring Harb Perspect Biol, 2*, 1-18.

55. Josten, N.J. & Huberman, A.D. in Invertebrate and Vertebrate Eye Development 229-259 (Elsevier Academic Press Inc, San Diego, 2010).

56. Cline, H.T. (2001). Dendritic arbor development and synaptogenesis. *Current Opinion in Neurobiology, 11*, 118-126.

57. Ball, S.L. et al. (2002). Role of the beta(2) subunit of voltage-dependent calcium channels in the retinal outer plexiform layer. *Investigative Ophthalmology & Visual Science 43*, 1595-1603.

58. Zhou, W., Horstick, E.J., Hirata, H. & Kuwada, J.Y. (2008). Identification and expression of voltage-gated calcium channel beta subunits in Zebrafish. *Developmental Dynamics, 237*, 3842-3852.

59. Ball, S.L. & Gregg, R.G. (2002). Using mutant mice to study the role of voltage-gated calcium channels in the retina. *Advances in Experimental Medicine and Biology, 514*, 439-450.

60. Peloquin, J.B., Rehak, R., Doering, C.J. & McRory, J.E. (2007). Functional analysis of congenital stationary night blindness type-2 CACNA1F mutations F742C, G1007R, and R1049W. *Neuroscience, 150*, 335-345.

61. Bayley, P.R. & Morgans, C.W. (2007). Rod bipolar cells and horizontal cells form displaced synaptic contacts with rods in the outer nuclear layer of the nob2 retina. *The Journal of Comparative Neurology, 500*, 286-298.

62. Raven, M.A. et al. (2008). Early afferent signaling in the outer plexiform layer regulates development of horizontal cell morphology. *The Journal of Comparative Neurology, 506*, 745-758.

63.   Wycisk, K.A. et al. (2006). Mutation in the Auxiliary Calcium-Channel Subunit CACNA2D4 Causes Autosomal Recessive Cone Dystrophy. *American journal of human genetics, 79*, 973-977.

64.   Haeseleer, F. et al. (2004). Essential role of Ca2+-binding protein 4, a Cav1.4 channel regulator, in photoreceptor synaptic function. *Nature Neuroscience, 7*, 1079-1087.

65.   Striessnig, J., Bolz, H.J. & Koschak, A. (2010). Channelopathies in Cav1.1, Cav1.3, and Cav1.4 voltage-gated L-type Ca2+ channels. *Pflügers Archiv, 460*, 361-374.

66.   Chang, B. et al. (2006). The nob2 mouse, a null mutation in Cacna1f: Anatomical and functional abnormalities in the outer retina and their consequences on ganglion cell visual responses. *Visual Neuroscience, 23*, 11-24.

67.   Mize, R.R., Graham, S.K. & Cork, R.J. (2002). Expression of the L-type calcium channel in the developing mouse visual system by use of immunocytochemistry. *Brain Res Dev Brain Res, 136*, 185-195.

68.   Ziburkus, J., Dilger, E.K., Lo, F.S. & Guido, W. (2009). LTD and LTP at the developing retinogeniculate synapse. *Journal of Neurophysiology, 102*, 3082-3090.

69.   Huba, R., Schneider, H. & Hofmann, H.D. (1992). Voltage-gated currents of putative GABAergic amacrine cells in primary cultures and in retinal slice preparations. *Brain Research, 577*, 10-18.

70.   Gleason, E., Mobbs, P., Nuccitelli, R. & Wilson, M. (1992). Development of functional calcium channels in cultured avian photoreceptors. *Visual Neuroscience, 8*, 315-327.

71.   Uchida, K. & Iuvone, P.M. (1999). Intracellular Ca2+ concentrations in cultured chicken photoreceptor cells: sustained elevation in depolarized cells and the role of dihydropyridine-sensitive Ca2+ channels. *Molecular Vision, 5*, 1.

72.   Kaneda, M., Ito, K., Morishima, Y., Shigematsu, Y. & Shimoda, Y. (2007). Characterization of voltage-gated ionic channels in cholinergic amacrine cells in the mouse retina. *Journal of Neurophysiology, 97*, 4225-4234.

73.   Guido, W. (2008). Refinement of the retinogeniculate pathway. *The Journal of Physiology, 586*, 4357-4362.

74.   Feller, M.B. (2009). Retinal waves are likely to instruct the formation of eye-specific retinogeniculate projections. *Neural Development, 4*, 24.

75.   Dolmetsch, R.E., Pajvani, U., Fife, K., Spotts, J.M. & Greenberg, M.E. (2001). Signaling to the Nucleus by an L-type Calcium Channel-Calmodulin Complex through the MAP Kinase Pathway. *Science, 294*, 333-339.

76.   Aguado, F. et al. (2009). The CREB/CREM Transcription Factors Negatively Regulate Early Synaptogenesis and Spontaneous Network Activity. *The Journal of Neuroscience, 29*, 328-333.

77.   Shi, L., Ko, M.L. & Ko, G.Y.-P. (2009). Rhythmic Expression of MicroRNA-26a Regulates the L-type Voltage-gated Calcium Channel alpha1C Subunit in Chicken Cone Photoreceptors. *Journal of Biological Chemistry, 284*, 25791-25803.

78.   Farrell, S.R., Raymond, I.D., Foote, M., Brecha, N.C. & Barnes, S. (2010). Modulation of Voltage-Gated Ion Channels in Rat Retinal Ganglion Cells Mediated by Somatostatin Receptor Subtype 4. *Journal of Neurophysiology, 104*, 1347-1354.

79.   Nieukoop, P.D. &.Faber, J. (ed.) (1994). *Normal Table of Xenopus Laevis (Daudin)*. New York: Garland Publishing, Inc.

80.   Wong, L.L. & Rapaport, D.H. (2009). Defining retinal progenitor cell competence in Xenopus laevis by clonal analysis. *Development, 136*, 1707-1715.

Chapter 4

UNRAVELING THE MULTIPLE FUNCTIONS OF THE A2/Δ SUBUNITS OF CALCIUM CHANNELS

Jesús García[§]

Department of Physiology and Biophysics, University of Illinois at Chicago, 835 South
Wolcott Avenue, MC 901, Chicago, IL, U. S.

Abstract

The $\alpha2/\delta$ subunits form part of voltage-dependent calcium channels and are considered accessory proteins. Initial studies of $\alpha2/\delta$ subunits revealed that they modify the voltage dependence, amplitude, and kinetics of calcium currents. $\alpha2/\delta$ subunits have also been identified as the receptor for gabapentin and pregabalin, drugs clinically used in the treatment of several neuropathic disorders. In addition to mediating therapeutical mechanisms, the involvement of $\alpha2/\delta$ subunits in diseases has been demonstrated in mouse models with mutations of the protein or alterations in expression levels. The structure of $\alpha2/\delta$ subunits and their pattern of expression suggested that these proteins may be participating in cellular functions separate from the action of calcium channels. Recently, it has been shown that $\alpha2/\delta$ subunits are more than just accessory subunits of calcium channels and that they are important for the trafficking of channels, attachment and migration of myoblasts, synaptogenesis, and other cellular processes. It is clear that $\alpha2/\delta$ subunits play many and important roles in the normal functioning of cells. This chapter reviews the influence of $\alpha2/\delta$ subunit in setting of the biophysical properties of calcium channels, the connection of $\alpha2/\delta$ with diseases, and the functions of $\alpha2/\delta$ that are now being unraveled.

The $\alpha2/\delta$ subunits are encoded by four genes

To date, four genes encoding $\alpha2/\delta$ subunits have been found. Of the genes that encode $\alpha2/\delta$ subunits, $\alpha2/\delta$-1 and $\alpha2/\delta$-2 are expressed in nearly every tissue (Angelotti & Hoffman, 1996; Klugbauer et al., 1999). The $\alpha2/\delta$-3 protein and mRNA have been found predominantly

---

[§] E-mail address: garmar@uic.edu

in mouse brain (Klugbauer et al., 1999; Barclay et al., 2001) while in humans only the RNA, not the protein, has been found in the heart (Gong et al., 2001). The α2/δ-4 protein is present in the pituitary, adrenal gland, colon, and fetal liver; RNA for α2/δ-4 was also detected in cardiac and skeletal muscle (Qin et al., 2002).

All of the α2/δ subunits experience alternative splicing giving rise to several isoforms. Five isoforms (*a* through *e*) of the α2/δ-1 result from alternative splicing in three regions. Expression of α2/δ-1 isoforms varies with development and is tissue-dependent. In immature mouse skeletal muscle three isoforms of α2/δ-1 can be found (*a*, *d*, and *e*) (Nabhani et al., 2005); as the animals develop, the levels of *d* and *e* isoforms are undetectable by day 30 and beyond, and the α2/δ-1a isoform is the only one expressed in adult muscle (Angelotti & Hofmann, 1996). Adult brain expresses only α2/δ-1b, heart expresses α2/δ-1c and α2/δ-1d, and smooth muscle tissues express α2/δ-1d and α2/δ-1e isoforms.

Two sequences of human α2/δ-2 that differ at the amino terminus have been found. However, it seems that only isoform I may be expressed and contain a potential signal sequence. Three different splice variants have been identified for α2/δ-2 isoform I (*a* though *c*) as a result of two alternative splicing regions (Hobom et al., 2000). The only α2/δ-2 splice variant detected in human heart was α2/δ-2a. High levels of expression of α2/δ-2 are also found in pancreas and skeletal muscle. Lower levels are found in liver, kidney, placenta, and several regions of the brain (Hobom et al., 2000).

Only one variant of the α2/δ-3 has been found. Expression of mRNA for α2/δ-3 has been found in cardiac and skeletal muscle but not at the protein level. It seems that α2/δ-3 is found only in the brain in the pyramidal layer of the hippocampus, caudate putamen, entorhinal complex, and cortex.

The amino and carboxy termini of α2/δ-4 show the highest variability and potentially would give rise to four variants (*a* through *d*). However, only variants *a* and *c* have a signal peptide sequence. The absence of a signal peptide in variants *b* and *d* suggest that these two variants would not reach the membrane. Abundant mRNA levels of α2/δ-4 were found in human heart and skeletal muscle. Lower levels were detected in brain, placenta, lung, liver, kidney, and pancreas. Although mRNA levels were high in heart, protein levels were not. The α2/δ-4 protein is found in the pituitary, adrenal gland, and brain (Qin et al., 2002).

The α2/δ subunit is part of voltage-dependent calcium channels

Voltage-dependent calcium channels are a heteromeric complex formed by four proteins, α1 or $Ca_vX.X$, α2/δ, β, and γ. Calcium channels were originally isolated from the transverse tubular system of skeletal muscle due to their high affinity to dihydropyridines (Curtis & Catterall, 1984; Borsotto et al., 1985; Flockerzi et al., 1986). In adult skeletal muscle, calcium channels (or dihydropyridine receptors, DHPRs) from t-tubules are directly involved in excitation-contraction coupling (ECC). Consequently, the relative contribution of each DHPR subunit to the ECC mechanism has been examined. The $Ca_v1.1$ subunit is the voltage sensor and contains the channel pore (Tanabe et al., 1988), while the other subunits of the DHPR regulate the activity of $Ca_v1.1$. The $Ca_v1.1$ and β subunits are essential components of the complex since muscle from dysgenic mice, which lack the $Ca_v1.1$ subunit, or from β-null mice have a complete loss of ECC and L-type calcium current (Tanabe et al., 1988; Gregg et al., 1996). In contrast, muscle from γ-null mice does not show changes in ECC and only modest effects on L-type calcium current (Freise et al., 2000). Recent evidence suggests that reducing the levels of α2/δ1 subunit by siRNA silencing has variable effects on the amplitude of I_{Ca-L} and calcium transients and release elicited by voltage clamp depolarizations, and is important for maintaining calcium transients during sustained depolarizations with KCl (Obermair et al., 2005; Garcia et al., 2008; Gach et al., 2008). These results suggest that although the α2/δ1 subunit is part of the DHPR, a skeletal muscle phenotype can be preserved *in vitro* in cells with decreased levels of this protein. Other than the studies with *in vitro* systems using siRNA, there is no more information available that could help conclusively delineate the role of α2/δ in EC coupling. If we were to draw a conclusion based on those studies, it would seem that α2/δ is not important for ECC and that its absence would not have any significant effect on muscle function, similarly to the γ subunit. However, and in contrast to the γ subunit, elucidating the role of the α2/δ1 subunit has proved difficult since targeting of α2/δ1 exon 8 in a knockout mouse model was lethal for embryonic development (J. Offord, Pfizer, personal communication). Due to the arrested development of these animals, a phenotype was impossible to be determined. It can only be speculated that a deleterious effect on the nervous system and/or other vital organ(s) might have been the cause for the death of the embryos. In addition, a detrimental effect on development of skeletal muscle cannot be discarded. In contrast, embryonic lethality was not observed in a more recent α2/δ1 knockout model in which exon 2 was disrupted (Fuller-Bicer et al., 2009). These mice had decreased basal contractility and relaxation of the heart and a decrease of L-type calcium current.

The arrangement of the α2/δ subunit in the membrane gives clues as to its interaction with α1

The α2/δ subunit is a dimer protein product of a single gene (Dejongh et al., 1990; Jay et al., 1991). Post-translation cleavage results in an α2 protein containing ~950 N-terminal residues and a δ protein with ~130 C-terminal residues. Initially, the hydrophobicity profile of the α2/δ peptide isolated from rabbit skeletal muscle revealed three putative transmembrane regions (Ellis et al., 1988). However, C-terminal deletions, which included only the putative δ transmembrane region, precluded the protein from membrane association (Gurnett et al., 1996) suggesting that the other two alpha helices do not pass through the membrane. Initial evidence suggested that portion of the δ peptide anchors the α2 protein in the plasma membrane via disulfide bridges in a cysteine-rich region (CysRx), and all of the α2 peptide and approximately half of the δ peptide remain extracellularly. However, recent evidence suggests that the majority of the α2/δ subunits are located outside the cells and attached to the membrane via glycosylphosphatidylinositol anchors bound to the δ peptide (Davies et al., 2010). This was the case for subunits α2/δ-1, α2/δ-2, and α2/δ-3; subunit α2/δ-4 was not examined. Although the precise protein topology is not definitive, possible sites where the α2/δ subunit interacts with the main pore-forming α1 subunit occur via the extracellular α2 peptide and/or within the membrane-spanning region if δ remains as a transmembrane peptide.

The effects of the α2/δ subunit on calcium currents depend on the experimental conditions

A myriad of studies has examined the function of the α2/δ subunit in the modulation of channel properties. The results depend greatly on the system and the nature of α2/δ and α1 subunits. The regions responsible for the modification of calcium channel activity have not been precisely elucidated. However, domain III of the α1 has been implicated in binding to predominantly extracellular α2 peptide (Gurnett *et al.*, 1997). In addition, the δ transmembrane portion of the protein may regulate the kinetics of channel activation and inactivation (Felix *et al.*, 1997) although the region of the α1 which may bind to or interact with this portion of the α2/δ peptide has not been identified. Experiments in heterologous expression systems coexpressing α2/δ with cardiac α1 in COS cells produced an increase in dihydropyridine binding sites, calcium current amplitude, and susceptibility to voltage-dependent inactivation (Wei et al., 1995; Bangalore et al., 1996; Felix et al., 1997). Furthermore, the α2/δ subunit has been shown to increase the levels of α1 in the plasma membrane, calcium currents, and single channel open probability (Po) in studies with *Xenopus* oocytes (Shistik et al., 1995). Therefore, at the functional level, α2/δ may participate in channel assembly and in enhancing the Po in expression systems. In contrast, reducing α2/δ-1 levels with siRNA in muscle cells resulted in an acceleration of the calcium current with dissimilar effects on current conductance. Reduction of α2/δ1 in non-fusing BC3H1 muscle cells had a small effect on current conductance (Obermair et al., 2005), while in transduced myoblasts the conductance was significantly reduced (Gach et al., 2008). Although a common feature of α2/δ1 seems to be setting up channel properties, the effect caused by this protein on calcium currents was opposite in expression systems (acceleration) and muscle cells (slowing). In addition, while in expression systems α2/δ1 augmented the amplitude of the current by increasing the level of α1 in the plasma membrane, there was no clear effect in muscle cells, suggesting that α2/δ-1 may not necessary for transport of α1 to the membrane. Interestingly, a mutation in another member of the α2/δ family (α2/δ2) resulted in reduced presence of several α1 subunits ($Ca_v1.2$, $Ca_v2.1$, $Ca_v2.2$) in tsA-201 cells (Canti et al., 2005). The lack of a consistent effect on calcium currents by the α2/δ1 subunit among the different systems suggests that this protein has other functions not yet elucidated.

While the association of α2/δ subunits with α1 subunits of the $Ca_v1.X$ and $Ca_v2.X$ families of calcium channels is clear, their association with $Ca_v3.X$ or low voltage-activated (LVA) channels in their native environment has remained uncertain. In heterologous systems, co-expression of α2/δ subunits with LVA channels ($Ca_v3.1$, $Ca_v3.2$) results in an increase in membrane localization of α1 subunits and amplitue of calcium currents (Dolphin et al., 1999; Hobom et al., 2000; Gao et al., 2000; Dubel et al., 2004). These results suggest that both subunits are able to interact with each other, at least when they are both over-expressed. An indication that native expression of α2/δ1 and $Ca_v3.2$ subunits leads to interaction of the two subunits has recently been provided by Thompson et al. (2011) using the osteocyte cell line MYO-L4. Membrane localization of $Ca_v3.2$ was reduced when α2/δ1 levels were also reduced with siRNA. The issue of whether HVA α1 subunits are always associated with α2/δ subunits or whether they can function without α2/δ subunits remains unresolved. Association of α2/δ with LVA α1 subunits is more complicated and seems to depend on the cell type used, whether the subunits are expressed natively or heterologously, and the combination of α2/δ and α1 subunits.

Role of the α2/δ1 subunit on calcium release

Because α2/δ subunits were first isolated from skeletal muscle, it was logical to examine their role in calcium release. Alden & Garcia (2002) provided the first study showing a possible participation of the α2/δ1 subunit in regulating calcium release in voltage-clamped primary myotubes using gabapentin. Gabapentin is a GABA derivative that binds to the α2/δ subunit, and preferentially to α2/δ1 (Wang et al., 1999) with high affinity. Gabapentin decreased the rate of calcium release (without modifying the maximum amplitude of the transient) in spite of causing an increase in the amount of immobilization-resistant charge movement, suggesting a dissociation of the DHPR functions in skeletal muscle EC coupling. In contrast, it has been reported that a reduction of α2/δ-1 levels with siRNA did not modify the amplitude or the voltage dependence of calcium transients (Obermair et al., 2005). However, there is a crucial difference between these later experiments and those of Alden & Garcia (2002). The experiments by Obermair et al (2005) were performed on a cell line derived from dysgenic muscle (GLT cells) expressing an exogenous α1 subunit. The problem with these rescued GLT cells is that calcium release is much slower than in normal myotubes and any reduction in the initial part of the calcium transient may have been easily missed. More recently, Gach et al. (2008) reported no difference in the amplitude of calcium transients produced by short electrical pulses in transformed myotubes. But they did not provide further details regarding the rate of calcium release. However, they did find that the α2/δ1 subunit is important for maintaining the amplitude of the transient during maintained depolarizations due to either KCl application or trains of electrical stimulation. These results suggest that the α2/δ1 subunit may have an effect on calcium handling in skeletal muscle cells, but its role in this mechanism is far from clear.

The structure of the α2/δ subunit provides evidence for a yet unidentified role of the α2/δ subunit in extracellular signaling

Previous knowledge about the function of the α2/δ subunit has been largely limited to the interaction of this protein with the α1 subunit and its effects on calcium currents. Interestingly, although the structure and pattern of expression of the α2/δ subunit during development are fascinating and provide significant clues about additional functions, their importance in signaling mechanisms different from EC coupling has been overlooked. Furthermore, even though the majority (or perhaps all) of the α2/δ subunit is located extracellularly, interaction of this protein with the extracellular millieu has not been studied in depth. The α2/δ subunit is highly glycosilated and it also contains other conserved domains thought to mediate cell signaling. The *N*-terminal half of α2 contains a von Willebrand A (VWA) domain (Bork & Rhode, 1991; Whittaker & Hynes, 2002). The VWA domain is found in cell adhesion and extracellular matrix proteins and, in many cases, it is possible that this domain is involved in protein-protein interactions requiring divalent cations. The VWA occurs most notably in integrins and extracellular matrix proteins. Included in the VWA domain of α2 is a metal ion-dependent adhesion site (MIDAS), which is probably important for mediating protein-protein interaction (Whittaker & Hynes, 2002), such as ligand-receptor binding. Mutations in the metal binding site of the MIDAS domain result in reduced trafficking of α1 subunits and consequently a decrease in I_{Ca} (Canti et al., 2005). Immediately adjacent to the C-terminal of the VWA domain (starting at residue 446 of α2), there are two Cache domains. These domains have been found only in this CAlcium channel subunit and prokaryotic CHEmotaxis receptors (Anantharaman & Aravind, 2000). The predicted membrane topology and the requirement of this domain in prokaryotes for sensing various molecules (i.e. amino acids, glucose, citrate), led Anantharaman & Aravind (2000) to suggest that the Cache domain is probably implicated in regulation of calcium channels by endogenous ligands. The presence of these conserved domains on α2 suggests that this subunit may be important for mediating signaling from extracellular sources. However, the contribution of the α2/δ subunit in processes other than EC coupling has not been previously realized.

To investigate whether the α2/δ1 subunit is important for cell-extracellular matrix interaction, Garcia et al. (2008) examined the effect of drastically reducing the levels of α2/δ1 protein in C2C12 cells using siRNA. Cells with low levels of α2/δ1 experienced reduced attachment, migration, and spreading compared to control cells or cells treated with control siRNA (Garcia et al., 2008). Interestingly, this new role of the α2/δ1 subunit is not limited to muscle cells since preliminary evidence from another laboratory suggests that α2/δ1 protein may interact with the heparan sulfate proteoglycan perlecan in bone cells and bone matrix (Thompson & Farach-Carson, personal communication).

Developmental differences in expression and localization of α2/δ-1 and α1 subunits indicate that these proteins do not always interact with each other

Further support for the idea proposing a new role for the α2/δ-1 subunit is provided by the fact that, during development of skeletal muscle, α2/δ-1 appears earlier than α1 and its levels remain high through the differentiation process. This has been demonstrated both at the mRNA (Varadi et al., 1989) and the protein levels (Morton & Froehner, 1989). In agreement with those results, Garcia et al. (2008) have found that the α2/δ-1 subunit localizes at the ends of cells with little or no association with α1 subunits in myotubes after 2 days of induction of fusion and differentiation. With longer times in differentiation medium, the localization of α2/δ-1 gradually becomes homogeneous until it co-localizes almost completely with α1. However, there was some α2/δ-1 subunit that did no co-localize with α1 even at later times in a number of myotubes. Interestingly, experiments performed in dysgenic muscle have shown that the α2/δ-1 subunit is normally expressed but that its distribution patterns are abnormal in the absence of α1 (Flucher et al., 1991). In dysgenic muscle cells, α2/δ-1 was found in the plasma membrane, around the nucleus, and in the transverse tubular membrane in a diffuse pattern. Accordingly, the localization pattern of α2/δ-1 in dysgenic cells closely resembles the findings in immature muscle where there is little or no α1 subunit to associate with α2/δ-1. These results strongly suggest that the assembly of the DHPR as a complex occurs at later times in development than it was previously believed and not as soon as the subunits are expressed (Flucher, 1991; 1996). They further suggest that the expression of α2/δ-1 subunit is regulated independently of the α1 protein and that the α2/δ-1 subunit is not only part of the DHPR but that it may interact with other cellular components, especially early in development.

The α2/δ subunits interact with other cellular components in addition to α1 subunits

The extracellular domains of the α2/δ subunit together with the results indicating that this protein is important for cell attachment and migration suggest that α2/δ-1 may be important for transmitting signals from the extracellular millieu to the inside of the cells. Association of α2/δ subunits with proteins other than subunits of calcium channels has been reported for isoforms 1 and 2. Studies using rabbit skeletal muscle cells in culture showed that the α2/δ-1 subunit co-localizes with the neural cell adhesion molecule (NCAM) and the cellular distribution of both these proteins changes concomitantly during myogenesis *in vitro*, although the functional significance of a putative physical interaction between these two proteins was not examined (Vandaele & Rieger, 1994). More recently, Eroglu et al. (2009) demonstrated that the α2/δ-1 subunit binds to thrombospondins in the central nervous system. Thrombospondins are proteins in the extracellular matrix secreted by astrocytes and mediate cell-cell and cell-matrix interactions (Christopherson et al., 2005). Interestingly, secretion of thrombospondins occurs during synaptogenesis in early postnatal life. The interaction between the α2/δ1 subunit and thrombospondins proved to be required for the formation of new synapses, most likely independently from calcium channel function. Similarly, the α2/δ-2 subunit has been shown to bind the lipid raft-associated proteins SLP-2, stomatin, and prohibitin and that this interaction may influence the functionality of calcium channels in the nervous system (Davies et al., 2006).

Recent evidence has shown that α2/δ-1 subunits are also able to assocate with a subunit of the ATP synthase complex, the ATP5b subunit, in developing skeletal muscle (Garcia, 2011). This association occurs at the level of the plasma and internal membranes. The interaction of α2/δ-1 and ATP5b proved to modulate calcium release at the time when α2/δ-1 and α1 subunits begin to form the calcium channel. All these examples indicate that the α2/δ subunits have functions independent from calcium channels.

Link between α2/δ subunits and disease

In skeletal muscle, involvement of the α1 subunit in the pathogenesis of hypokalemic periodic paralysis (Ptacek et al., 1994; Jurkatt-Rott et al., 1994) and malignant hyperthermia susceptibility (MHS) type 5 has been established (Monnier et al., 1997; Jurkatt-Rott et al., 2000). In contrast, the participation of the α2/δ-1 subunit in muscle pathology is more tenuous and only a genetic linkage has been established between the locus of this subunit and MHS (Iles et al., 1994). To date no mutations have been identified in the α2/δ-1 subunit.

A mouse model of absence epilepsy and cerebellar ataxia is the mutant mouse *ducky*. These mice are smaller in size and present abnormalities of the dendritic tree of Purkinje cells. The underlying cause of this defect is a mutation in the Cacna2d2 gene with a head to tail duplication of exons 2-39 (Barclay et al., 2001). The duplication results in two truncated transcripts and dissappearance of the $\alpha2/\delta$-2 subunit. Purkinje cells from these mice display a significantly reduced calcium current and spontaneous firing (Barclay et al., 2001). Mutations in the human CACNA2D2 gene have not been reported.

REFERENCES

Alden, K.J. & García, J. (2002). Dissociation of charge movement from calcium release and calcium current in skeletal myotubes by gabapentin. *Am. J. Physiol. Cell Physiol.* 283:C941-C949.

Anantharaman, V. & Aravind, L. (2000). Cache – a signaling domain common to Ca2+ channel subunits and a class of prokaryotic chemotaxic receptors. *Trends Biochem. Sci* 25:535-5367.

Angelotti, T. & Hofmann, F. (1996). Tissue-specific expression of splice variants of the mouse voltage-gated calcium channel $\alpha2/\delta1$ subunit. FEBS Lett. 397:331-337.

Bangalore, R., Mehrke, G., Gingrich, K., Hofmann, F. and Kass, R.S. (1996) Influence of L-type Ca channel $\alpha2/\delta$-subunit on ionic and gating current in transiently transfected HEK 293 cells. *Am. J. Physiol.* 270:H1521-H1528.

Barclay, J. Balaguero, N., Mione, M., Ackerman, S.L., Letts, V.A., Brodbeck, J., Canti, C., Meir, A., Page, K.M., Kusumi, K., Perez-Reyes, E., Lander, E.S., Frankel, W.N., Gardiner, R.M., Dolphin, A.C., Rees, M. (2001) Ducky mouse phenotype of epilepsy and ataxia is associated with mutations in the Cacna2d2 gene and decreased calcium channel current in cerebellar Purkinje cells. *J. Neurosci.* 21, 6095–6104

Bork, P & Rohde, K. (1991). More von Willebrand factor type A domains? Sequence similarities with malaria thrombospondin-related anonymous protein, dihydropyridine sensitive calcium channel and inter-alpha-trypsin inhibitor. *Biochem. J.* 279:908-910.

Borsotto, M., Barhanin, J., Fosset, M., and Lazdunski, M. (1985) The 1,4-dihydropyridine receptor associated with the skeletal muscle voltage-dependent Ca2+ channel. Purification and subunit composition. *J. Biol. Chem.* 260,14255-14263.

Canti, C., Nieto-Rostro, M., Foucault, I., Heblich, F., Wratten, J., Richards, M.W., Hendrich, J., Douglas, L., Page, K.M., Davies, A. & Dolphin, A.C. (2005). The metal-ion-dependent adhesion site in the VonW illebrand factor-A domain of alpha2delta subunits is key to trafficking voltage-gated Ca2+ channels. *Proc. Natl. Acad. Sci. USA,* 102:11230-11235.

Christopherson, K.S., Ullian, E.M., Stokes, C.C., Mullowney, C.E., Hell, J.W., Agah, A., Lawler, J., Mosher, D.F., Bornstein, P., Barres, B.A. (2005). Thrombospondins are astrocyte-secreted proteins that promote CNS synaptogenesis. *Cell* 120:421-433.

Curtis, B. M., and Catterall, W. A. (1984). Purification of the calcium antagonist receptor of the voltage-sensitive calcium channel from skeletal muscle transverse tubules. *Biochemistry* 23:2113-2118.

Davies, A., Douglas, L., Hendrich, J., Wratten, J., Tran Van Minh, A., Foucault, I., Koch, D., Pratt, W.S., Saibil, H.R. and Dolphin, A.C. (2006). The calcium channel α2δ-2 subunit partitions with CaV2.1 into lipid rafts in cerebellum: implications for localization and function. *J. Neurosci.* 26: 8748-8757.

Davies, A., Kadurin, I., Alvarez-Laviada, A., Douglas, L., Nieto-Rostro, M., Bauer, C.S., Pratt, W.S., Dolphin, A.C. (2010). The α2δ subunits of voltage-gated calcium channels form GPI-anchored proteins, a posttranslational modification essential for function. *Proc Natl Acad Sci U S A.*107:1654-1659.

De Jongh, K. S., Warner, C. and Catterall, W. A. (1990). Subunits of purified calcium channels. Alpha 2 and delta are encoded by the same gene. *J Biol Chem* 265:14738-14741.

Dolphin, A.C. Wyatt, C. N., Richards, J., Beattie, R. E., Craig, P., Lee, J.-H., Cribbs, L. L., Volsen, S. G., and Perez-Reyes, E. (1999) The effect of α2-δ and other accessory subunits on expression and properties of the calcium channel α1G. *J. Physiol.* 519:35–45.

Dubel, S.J., Altier, C., Chaumont, S., Lory, P., Bourinet, E., and Nargeot, J. (2004) Plasma membrane expression of T-type calcium channel α1 subunits is modulated by high voltage-activated auxiliary subunits. *J. Biol. Chem.* 279:29263–29269.

Ellis, S. B., Williams, M. E., Ways, N. R., Brenner, R., Sharp, A. H., Leung, A. T., Campbell, K. P., McKenna, E., Koch, W. J., Hui, A., Schwartz, A. and Harpold, M. M. (1988). Sequence and expression of mRNAs encoding the alpha 1 and alpha 2 subunits of a DHP-sensitive calcium channel. *Science* 241:1661-1664.

Eroglu, C., Allen, N.J., Susman, M.W., O'Rourke, N.A., Park, C.Y., Ozkan, E., Chakraborty, C., Mulinyawe, S.B., Annis, D.S., Huberman, A.D., Green, E.M., Lawler, J., Dolmetsch, R., Garcia, K.C., Smith, S.J., Luo, D., Rosenthal, A., Mosher, D.F., Barres, B.A. (2009). Gabapentin receptor α2δ-1 is a neuronal thrombospondin receptor responsible for excitatory CNS synaptogenesis. *Cell* 139:380-392.

Felix, R., Gurnett, C. A., De Waard, M. and Campbell, K. P. (1997). Dissection of functional domains of the voltage-dependent Ca^{2+} channel alpha2delta subunit. *J Neurosci* 17:6884-6891.

Flockerzi, V., Oeken, H.-J., Hofmann, F., Pelzer, D., Cavalie, A., and Trautwein, W. (1986). Purified dihydropyridine-binding site from skeletal muscle t-tubules is a functional calcium channel. *Nature* 323,66-68.

Flucher, B.E., Phillips, J.L. & Powell, J.A. (1991). Dihydropyridine receptor α subunits in normal and dysgenic muscle in vitro: expression of α1 is required for proper targeting and distribution of α2. *J. Cell Biol.* 115:1345-1356.

Flucher, B.E. & Franzini-Armstrong, C. (1996). Formation of junctions involved in excitation-contraction coupling in skeletal and cardiac muscle. *Proc. Natl. Acad. Sci. USA* 93:8101-8106.

Freise, D., Held, B., Wissenbach, U., Pfeifer, A., Trost, C., Himmerkus, N., Schweig, U., Freichel, M., Biel, M., Hofmann, F., Hoth, M. & Flockerzi, V. (2000). Absence of the γ subunit of the skeletal muscle dihydropyridine receptor increases L-type Ca^{2+} currents and alters channel inactivation properties. *J. Biol. Chem.* 275:14476-14481.

Fuller-Bicer, G.A., Varadi, G., Koch, S.E., Ishii, M., Bodi, I., Kadeer, N., Muth, J.N., Mikala, G., Petrashevskaya, N.N., Jordan, M.A., Zhang, S.P., Qin, N., Flores, C.M., Isaacsohn, I., Varadi, M., Mori, Y., Jones, W.K. & Schwartz, A. (2009). Targeted disruption of the voltage-dependent calcium channel α2/δ-1-subunit. *Am. J. Physiol. Heart Circ. Physiol.* 297:H117-H124.

Gach, M.P., Cherednichenko, G., Haarmann, C., Lopez, J.R., Beam, K.G., Pessah, I.N., Franzini-Armstrong, C. & Allen, P.D. (2008). α2δ1 dihydropyridine receptor subunit is a critical element for excitation-coupled calcium entry but not for formation of tetrads in skeletal muscle. *Biophys. J.* 94:3023-3034.

Gao, B. Yoshitaka, S., Maximov, A., Saad, M., Forgacs, E., Latif, F.,Wei, M. H., Lerman, M., Lee, J.-H., Perez-Reyes, E., Bezprozvanny, I., and Minna, J. D. (2000) Functional properties of a new voltage-dependent calcium channel α2δ auxiliary subunit gene (CACNA2D2). *J. Biol. Chem.* 275:12237–12242.

García, J. (2011). The calcium channel α2/δ1 subunit interacts with ATP5b in the plasma membrane of developing muscle cells. *Am. J. Physiol. Cell. Physiol.* In Press.

Garcia, K., Nabhani, T., Garcia, J. (2008). The Calcium Channel α2/δ1 Subunit is Involved in Extracellular Signaling. *J. Physiol. (London).* 586.3:727-738.

Gong, H. C., Hang, J., Kohler,W., Li, L., and Su, T.-Z. (2001). Tissue-specific expression and gabapentin-binding properties of calcium channel α2δ subunit subtypes. *J. Membr. Biol.* 184:35–43.

Gregg RG, Messing A, Strube C, Beurg M, Moss R, Behan M, Sukhareva M, Haynes S, Powell JA, Coronado R & Powers PA (1996). Absence of the β subunit (cchb1) of the skeletal muscle dihydropyridine receptor alters expression of the α1 subunit and eliminates excitation-contraction coupling. *Proc. Natl. Acad. Sci. U S A.* 93:13961-13966.

Gurnett, C. A., De Waard, M. and Campbell, K. P. (1996). Dual function of the voltage-dependent Ca^{2+} channel alpha 2 delta subunit in current stimulation and subunit interaction. *Neuron* 16:431-440.

Hobom, M., Dai, S., Marais, E., Lacinova, L., Hofmann, F., Klugbauer, N. (2000). Neuronal distribution and functional characterization of the calcium channel α2δ-2 subunit. *Eur. J. Neurosci.* 12, 1217–1226.

Iles, D., Lehmann-Horn, F., Deufel, T., Scherer, S. W., Tsui, L.-C., Olde Weghuis, D., Suijkerbuijk, R. F., Heytens, L., Mikala, G., Schwarts, A., Ellis, F.R., Stewart, A.D., & B. Wieringa. (1994). Localization of the gene encoding the α_2/δ-subunits of the L-type voltage-dependent calcium channel to chromosome 7q and segregation of flanking markers in malignant hyperthermia susceptible families. *Hum. Mol. Genet.* 3:969-975.

Jay, S. D., Sharp, A. H., Kahl, S. D., Vedvick, T. S., Harpold, M. M. and Campbell, K. P. (1991). Structural characterization of the dihydropyridine-sensitive calcium channel alpha 2-subunit and the associated delta peptides. *J. Biol. Chem.* 266:3287-3293.

Jurkat-Rott, K., Lehmann-Horn, F., Elbaz, A., Heine, R., Gregg, R. G., Hogan, K., Powers, P., Lapie, P., Vale-Santos, J. E., Wwissenbach, J. & Fontaine, B. (1994). A calcium channel mutation causing hypokalemic periodic paralysis. *Hum. Mol. Genet.* 3:1415-1419.

Jurkat-Rott, K., McCarthy, T.V., & Lehmann-Horn, F. (2000) Genetics and pathogenesis of malignant hyperthermia. *Muscle Nerve* 23:4-17.

Klugbauer, N., Lacinova, L., Marais, E., Hobom, M., and Hofmann, F. (1999). Molecular diversity of the calcium channel $\alpha2\delta$ subunit. *J. Neurosci.* 19, 684–691.

Monnier, N., Procaccio, V., Stieglitz, P., & Lunardi, J. (1997). Malignant-hyperthermia susceptibility is associated with a mutation of the α_1-subunit of the human dihydropyridine-sensitive L-type voltage-dependent calcium-channel receptor in skeletal muscle. *Am. J. Hum. Genet.* 60:1316-1325.

Morton, M. & Froehner, S. (1989). The $\alpha1$ and $\alpha2$ polypeptides of the dihydropyridine-sensitive calcium channel differ in developmental expression and tissue distribution. *Neuron.* 2:1499-1506.

Nabhani, T., Shah, T. & García, J. (2005). Skeletal Muscle Cells Express Different Isoforms of the Calcium Channel $\alpha2/\delta$ Subunit. *Cell Bioch. Biophys.* 42:13-20.

Obermair, G.J., Kugler, G., Baumgartner, S., Tuluc, P., Grabner, M. & Flucher, B.E. (2005). The Ca^{2+} channel $\alpha2\delta$-1 subunit determines Ca^{2+} current kinetics in skeletal muscle but not targeting of $\alpha1s$ or excitation-contraction coupling. *J. Biol. Chem.* 280:2229-2237.

Ptacek, L.J., Tawil, R., Griggs, R.C., Engel, A., Layzer, R.B., Kwiecinski, H., McMamis P. G., Santiago L., Moore, M., Fouad, G., Bradley, P. & Leppert. M. F. (1994). Dihydropyridine receptor mutations cause hypokalemic periodic paralysis. *Cell* 77:863-868.

Qin, N., Yagel, S., Momplaisir, M.-L., Codd, E. E., and D'Andrea, M. R. (2002). Molecular cloning and characterization of the human voltage-gated calcium channel $\alpha(2)\delta$-4 subunit. *Mol. Pharmacol.* 62:485-496.

Shistik, E., Ivanina, T., Puri, T., Hosey, M. and Dascal, N. (1995). Ca^{2+} current enhancement by alpha 2/delta and beta subunits in Xenopus oocytes: contribution of changes in channel gating and alpha 1 protein level. *J Physiol (Lond)* 489:55-62.

Tanabe, T., Beam, K.G., Powell, J.A. and Numa, S. (1988) Restoration of excitation-contraction coupling and slow calcium current in dysgenic muscle by dihydropyridine receptor complementary DNA. *Nature* 328:313-318.

Thompson, W.R., Majid, A.S., Czymmek, K.J., Ruff, A.L., Garcia, J., Duncan, R.L., Farach-Carson, M.C. (2011). Association of the $\alpha2\delta1$ Subunit with Cav3.2 Enhances Membrane Expression and Regulates Mechanically Induced ATP Release in MLO-Y4 Osteocytes. *J. Bone Mineral Res.* Submitted.

Thurlow, R.J., Brown, J.P., Gee, N.S., Hill, D.R. & Woodruff, G.N. (1993). [^{3}H]gabapentin may label a system-L-like neutral amino acid carrier in brain. *Eur J Pharmacol.* 247:341-345.

Vandaele, S.F. & Rieger, F. (1994). Co-localization of 1,4-dihydropyridine receptor $\alpha2/\delta$ subunit and N-CAM during early myogenesis in vitro. *J. Cell Science* 107:1217-1227.

Varadi G, Orlowski J & Schwartz A (1989). Developmental regulation of expression of the alpha 1 and alpha 2 subunits mRNAs of the voltage-dependent calcium channel in a differentiating myogenic cell line. *FEBS Lett.* 250:515-518.

Wang, M., Offord, J., Oxender, D. L. and Su, T. Z. (1999). Structural requirement of the calcium-channel subunit alpha2delta for gabapentin binding. *Biochem. J.* 342:313-320.

Wei, X., Pan, S., Lang, W., Kim, H., Schneider, T., Perez-Reyes, E. and Birnbaumer, L. (1995). Molecular determinants of cardiac Ca^{2+} channel pharmacology. Subunit requirement for the high affinity and allosteric regulation of dihydropyridine binding. *J. Biol. Chem.* 270:27106-27111.

Whittaker, C.A. & Hynes, R.O. (2002). Distribution and evolution of von Willebrand/Integrin A domains: widely dispersed domains with roles in cell adhesion and elsewhere. *Mol. Biol. Cell* 13:3369-3387.

In: Calcium Channels: Properties, Functions and Regulation   ISBN 978-1-61470-232-0
Editor: Mark R. Figgins   © 2012 Nova Science Publishers, Inc.

Chapter 5

MESOSCOPIC SIMULATION OF SUBCELLULAR CALCIUM MICRODOMAINS AND CALCIUM-REGULATED CALCIUM CHANNELS

Nicolas Wieder,[*] Rainer H. A. Fink and Frederic von Wegner[†]

Abstract

In this chapter we present a stochastically exact, full Markovian model of localized, subcellular calcium dynamics in femtoliter volumes based on the Gillespie algorithm. We show that a stochastic approach is necessary to accurately describe the events that constitute the relevant steps of subcellular signal processing in many cell types, and especially in excitable tissues. We discuss how different buffer systems shape the stochastic properties of the local calcium signal on several time scales. Using the chemical Langevin equation (CLE) as an approximation to the solution computed with Gillespie's algorithm, the local calcium signal can be treated as a colored noise input to downstream calcium-dependent signalling cascades. Importantly, the colored noise approximation achieved by the CLE approach allows for accelerated simulations.

To exemplify the procedure we will include the calcium regulated IP_3R calcium channel in the simulations to produce a local positive feedback loop. This system represents the basic mechanism responsible for elementary calcium release events (blips, puffs, sparks) as well as for global calcium waves and oscillations. The inherent stochasticity of localized calcium dependent signalling is illustrated by including calmodulin and calmodulin dependent protein kinase II α (CaMKIIα) in the model and it is shown that individual time courses of activated CaMKII α display a large variability that has to be taken into account when investigating downstream signalling events such as CaMKIIα dependent protein phosphorylation-dephosphorylation switches.

Keywords: single voxel stochastic simulation, Gillespie algorithm, calcium, calcium channels, microdomains.

---

[*]E-mail address: nwieder@ix.urz.uni-heidelberg.de
[†]E-mail address: frederic.wegner@urz.uni-heidelberg.de

1. Introduction

Calcium ions (Ca^{2+}) implement a highly versatile intracellular signalling system in virtually all cell types. The intracellular calcium concentration $\left[Ca^{2+}\right]$ is tightly regulated by a number of conserved mechanisms distributed across various subcellular compartments [?]. In contractile cells such as skeletal muscle or cardiac myocytes for instance, the rapid global increase of the intracellular Ca^{2+} concentration initiates motor protein interaction and at the same time, activation of Ca^{2+}-sensitive metabolic pathways adapts the cellular metabolism to the increased energetic demand [?]. In neurons, localized Ca^{2+} elevations in the dendritic tree of the cell allow synaptic plasticity to occur in a highly differentiated, input specific manner [?]. Apart from the excitable cell types mentioned, ubiquitous Ca^{2+} regulated cellular processes comprise the regulation of the plasma membrane potential, cellular motility, differentiation and cell cycle control, and metabolic activity [?]. This regulatory function is mediated by a large number of Ca^{2+} sensitive cellular proteins such as ion channels, protein kinases and phosphatases, enzymes and transcription factors.

Due to the involvement of Ca^{2+} in numerous central signalling pathways, the subcellular Ca^{2+} dynamics are of central interest to system biologists, even when the primary interest is not Ca^{2+} dynamics per se. Given the complexity of these models even before the inclusion of a calcium subsystem, two main questions arise. First, how much detail is necessary to simulate realistic Ca^{2+} fluctuations, and second, which modeling strategy (deterministic or stochastic) should be followed. In this article, we argue in favor of a fully stochastic simulation of the Ca^{2+} dynamics rather than a deterministic reaction rate system, in particular when small volumina and low copy numbers of signalling molecules are considered. The reason for this approach is that in past years we have learned about the relevance of stochastic effects in 'small' systems, especially in information processing systems such as molecular signalling cascades [?, ?].

We present the algorithmic details of how to implement small, but generic Ca^{2+} subsystems including buffers, Ca^{2+} regulated ion channels (IP_3R) and diffusion events within the framework of Gillespie's exact stochastic simulation algorithm [?, ?, ?]. Stochastic properties of the model systems are quantified in the time and frequency domains to illustrate the connection to standard methods of stochastic analysis. We investigate simulation volumes of ¡ 1 fl (1 μm^3) and show that the stochasticity of local Ca^{2+} dynamics is relevant on these scales. The variability in the activation profiles of downstream calcium sensitive signalling molecules is illustrated for the case of the calcium/calmodulin dependent protein kinase CaMKIIα. Finally, we discuss the chemical Langevin equation as an elegant and very fast alternative to the exact stochastic simulation algorithm that, under proper conditions, yields realistic Ca^{2+} time courses at a low computational cost [?, ?]. The (sub-)femtoliter scale investigated here represents the size of individual organelles (e.g. mitochondria, nuclei) and of macromolecular complexes such as intracellular ion channel clusters. Furthermore, this scale represents exactly the volumes recorded by current confocal microscopes from which data for quantitative fluorescence analysis are obtained.

2. Simulation of Chemical Reactions

2.1. The Gillespie Algorithm

We are using an extension of Gillespie's algorithm for the stochastic simulation of a defined set of chemical reactions. The Gillespie approach is based on the idea of transforming the predefined macroscopic reaction rate constants into time- and state-dependent probabilities [?]. The model system is a multivariate Markov process whose evolution is fully described by an initial state, a sequence of waiting times and state-dependent (time inhomogeneous) transition probabilities. The approach proves to be very powerful as any process described by a rate constant can be integrated in the model in a straightforward way. For instance, transitions between functional states of ion channel proteins, ion permeation through a channel pore and importantly, diffusion events can be included in exact stochastic simulation algorithms [?, ?, ?]. We now give a short summary of the original Gillespie algorithm [?].

We assume a well stirred chemical system whose state is represented by the copy number of each reactant, i.e. we ignore the velocity and position coordinates of individual molecules. This assumption is based on the fact that nonreactive molecular collisions that lead to a mixing of the system are far more probable than the occurrence of reactive events. This leads to uniformly randomized positions and thermally randomized velocities throughout the volume (Maxwell-Boltzmann distribution) [?]. The model system is characterized by a set of properties and a set of events. In the following, we define an event as a biophysical process which is characterized by a rate constant. System properties are the simulation volume Ω and the set S of N reaction Species $S = \{S_1, ..., S_N\}$. The state vector $\mathbf{x}(t) = (x_1(t), ..., x_N(t))$ represents the number of molecules of each species S_k, $k = 1, ..., N$ in the volume Ω.

Events define the possible transitions between different system states. The action of an event R_j on the system state, i.e. the induced transition, is implemented via the state change vector $\mathbf{v}_j = (d_{j1}, ..., d_{jN})$, $j = 1, ..., M$ where component d_{ji} defines the change in the number of the molecular species S_i due to reaction R_j. The reaction propensity a_j of different events is calculated based on the (stochastic) reaction rate c_j and the stoichiometric coefficients of the underlying process. In the following, we exemplify the calculation of the relevant event propensities:

- **Reaction events:** $a_j(\mathbf{x}, t) = c_j * h_j(\mathbf{x}, t)$ where $h_j(\mathbf{x}, t)$ is derived from the stoichiometric coefficients of the reaction and denotes the number of possible molecular combinations available for reaction R_j at time t.

  For the mono- and bimolecular reactions in the models presented here, the stochastic rate constants c_j are calculated from the macroscopic reaction rate constants k_j as $c_j = k_j$ for monomolecular reactions and as $c_j = k_j / \Omega$ for bimolecular reactions given the system volume Ω.

- **Diffusion events:** $a_i(\mathbf{x}, t) = c_i * x_i(t)$ where $x_i(t)$ denotes the copy number of the diffusible species S_i at time t and $c_i = D_i / A$ where D_i denotes the diffusion coefficient of species S_i and A denotes the area of diffusion.

- **Channel gating:** here, we consider an IP_3R calcium channel. The channel protein is a homotetramer and the (discrete) channel state is given by the combination of its four subunit states. The subunit transition rates are Ca^{2+} and IP_3 dependent and are simulated as mono- and bimolecular reactions according to the DeYoung-Keizer model [?].

- **Channel conductance:** Ion flux through a channel pore is a stochastic process. Here, we represent the channel flux as a Poisson process with a constant 'reaction' rate, i.e. a flux rate proportional to the channel conductance in order to yield the correct mean current. We get $c_j = I_{ch}/(e * z_{ion})$ where I_{ch} denotes the channel current in $[C/s]$, e is the elementary charge, and z_{ion} is the charge of the transported ion.

Sample paths generated by the Gillespie algorithm are samples of a multivariate Markov process. This means that, given a system state $\mathbf{x}(t)$ and a waiting time τ for the next reaction, the state $\mathbf{x}(t+\tau)$ only depends on $\mathbf{x}(t)$. Transitions between system states are performed by generating independent, exponentially distributed waiting times τ_j for each possible event R_j (according to the event propensities), and choosing the event with the minimum waiting time. In the software, τ-values are stored in a waiting-time queue. The computational procedure is the same for all event types R_j. It uses the current event propensity $a_j(\mathbf{x},t)$ and a uniformly distributed random variable $r \sim U_{[0,1]}$:

$$\tau = \frac{-\ln(r)}{a_j(\mathbf{x},t)} \tag{1}$$

Equation (1) leads to exponentially distributed waiting times τ which are characteristic for Markovian processes. After selection of the event R_j with the smallest τ-value τ_j, the current simulation time t is advanced by $t \leftarrow t + \tau_j$ and the system state is changed to $\mathbf{x}(t) \leftarrow \mathbf{x}(t) + \mathbf{v}_j$.

To increase the computational performance, the Gibson-Bruck algorithm introduced a dependency graph which leads to an efficient update rule where only those reaction propensities are updated that are influenced by the last reaction realized [?]. A reaction R_b depends on a reaction R_a if one of the educts of R_b is changed by reaction R_a.

2.2. Multivoxel simulations and boundary conditions

As Gillespie's algorithm assumes a well mixed simulation volume Ω, this volume cannot be chosen arbitrarily large. Therefore, depending on the desired simulation volume, it may become necessary to simulate multivoxel volumes. In multivoxel simulations, substance transport between adjacent voxels is realized by explicitly modelling diffusion events. It is observed that diffusion coefficients of small molecules, as those considered here, lead to diffusion event propensities that can be several orders of magnitude larger than chemical reaction propensities. However, the gain of precision and spatial detail of multi-voxel simulation leads to a significant increase of computation time. Previous studies suggest that single voxel simulations with volumes between 0.125 fl and 1 fl appear to be a good compromise between accuracy and computation time [?].

Within the single- and multi-voxel approaches, boundary conditions are implemented by assigning special rules to events occurring in boundary voxels (in the single voxel case, the unique voxel apparently is a boundary voxel). Using a cubic voxel geometry, each voxel has six neighbors where diffusible species can be transported to. When a diffusive event occurs in a boundary voxel, any standard boundary condition (absorption, reflection, periodic) can be implemented easily by defining the appropriate neighboring voxel.

In the single voxel case, we use a 'constant pool' assumption, i.e. we define a region outside the simulation volume, and set all diffusion propensities in the outside region to the equilibrium values of the system inside. Thus, in equilibrium, the net diffusive flow vanishes for all molecular species. Moreover, in case of channel mediated Ca^{2+} release, the system returns to the equilibrium concentrations as the extra amount of released Ca^{2+} ions leaves the volume by diffusion.

2.3. Calcium buffering

Calcium ions participate in a number of different buffer reactions with both, mobile and immobile buffers. Furthermore, many calcium channel types have regulatory sites that can bind Ca^{2+} ions, and can therefore also be considered calcium buffers. Buffer reactions are reversible and therefore lead to pairs of second order association and first order dissociation reactions:

$$Ca^{2+} + B \xrightarrow{k_j^+} CaB \tag{2}$$

$$CaB \xrightarrow{k_j^-} Ca^{2+} + B \tag{3}$$

The reaction rate constants k_j^+ and and k_j^- describe association and dissociation reactions, respectively. All numerical values used in the following examples are summarized in Tables 1 and 2.

2.4. Simulation of the IP_3R calcium channel

Intracellular calcium channels play an important role in the cellular response to external stimuli, in particular to receptor mediated signalling events exerted by hormones or neurotransmitters. The two most important intracellular calcium channel types are the inositol 1,4,5-trisphosphate receptor (IP_3R) and the ryanodine receptor (RyR) channels. Their gating (open-close) dynamics are regulated by a large number of biochemical mechanisms. Interestingly, Ca^{2+} ions themselves can both, activate and inhibit these channels in a concentration dependent manner. The IP_3R calcium channel is an ubiquitous and well characterized channel. In the models presented here, we use the classical DeYoung-Keizer model [?]. In this model, the functional state of the homotetrameric channel is determined by the four subunit states. The open configuration is assumed when at least 3 of the 4 subunits are in the open conformation. Each subunit can undergo modification on three different binding sites. An activating IP_3 site, an activating Ca^{2+} site and an inhibiting Ca^{2+} site. The gating scheme is shown in Figure 1. The indices of each state indicate which binding sites are occupied by their reactants, the first index represents the IP_3 site, the second index the activating Ca^{2+}- and the third index the inhibiting Ca^{2+} binding site. Transitions between different subunit states are simulated as first order reactions for the dissociation and as second order reactions for association events respectively. Calcium ion conductance is simulated as a zero-order reaction event with a constant rate that yields the correct mean current.

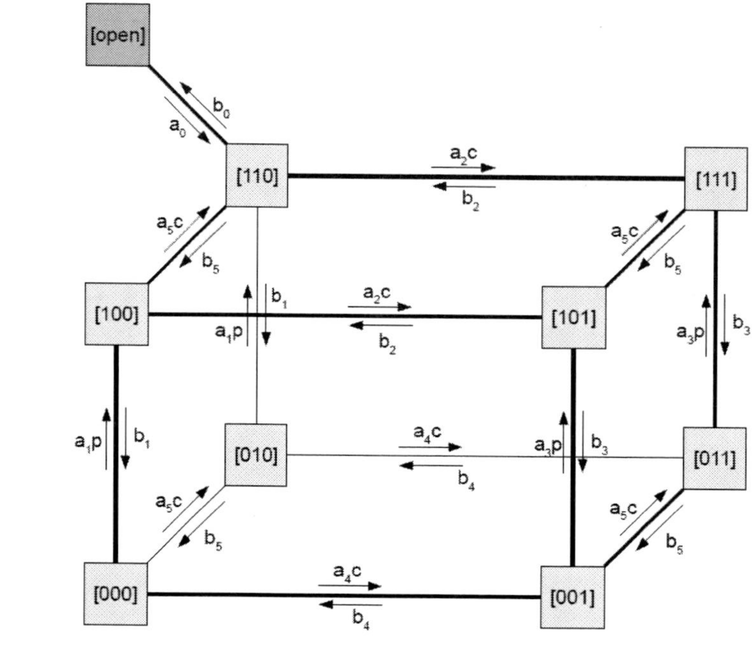

Figure 1. The DeYoung-Keizer gating model of the IP_3R calcium channel [?]: the four subunits of the IP_3R channel can undergo state transitions upon reactions with IP_3 and Ca^{2+}. The illustration summarizes the possible transition patterns and the corresponding rate constants. The three bit indices of each subunit represent the binding sites, the first index denotes IP_3 binding, the second index the activating Ca^{2+} and the third index the inactivating Ca^{2+} binding site [?].

2.5. Algorithmic Realization

0. Initialization: to initialize the simulation system, set the initial state $x_i(t_0)$ to the number of molecules S_i expected in equlibrium and calculate the waiting times τ_j, $j = 1, ..., M$ for each event R_j.

1. Event selection: select the event R_j that minimizes the associated τ-value.

2. Update time and state: increment the system time $t \leftarrow t + \tau_j$ and change the state vector according to $R_j : \mathbf{x}(t + \tau_j) = \mathbf{x}(t) + \mathbf{v}_j$.

3. Exit condition: if $(t + \tau_j > T_{max})$ then *Exit*, where T_{max} is the maximum simulation time.

4. Update: recalculate τ-values of related events as stored in the dependency graph.

  5. *goto step 1*

3.   Simulation Results

3.1.   Calcium fluctuations at different simulation volumes

To examine the influence of the system size, we performed simulations at different volumes. Figure 2 illustrates sample paths of the free Ca^{2+} concentration for different voxel sizes. The resulting time-averaged distribution of Ca^{2+} concentrations is shown in the right column as a histogram. Simulations were run in the presence of three Ca^{2+} buffers and without calcium channels. The reduced system containing only Ca^{2+}, a stationary and a mobile buffer species B_s, B_m as well as an external calcium dye B_{dye} is better suited to emphasize the influence of different simulation volumes in absence of the feedback loop induced by the channel.

The simulation results are shown for a mean $[Ca^{2+}]$ of 100 nM and the volumes $(1\,\mu m)^3 = 1$ fl (upper row), $(0.5\,\mu m)^3 = 0.125$ fl (middle row) and $(0.25\,\mu m)^3 = 0.015625$ fl (lower row). The rate constants for buffer reactions and diffusion events are identical to [?] and are given in Table 1 in the Appendix. The total buffer concentrations are $[B_s]_{total} = 300\,\mu M$, $[B_m]_{total} = 80\,\mu M$ and $[B_{dye}]_{total} = 40\,\mu M$ [?]. The histograms show clearly that small simulation volumes lead to skewed (Poissonian) concentration distributions, whereas larger volumes lead to nearly normally distributed calcium concentrations. This observation is later used to justify the introduction of the chemical Langevin equation as a fast approximation to Gillespie's algorithm. Clearly, the relative amount of fluctuation is larger for small volumes. The Ca^{2+} concentrations (mean $\pm$ standard deviation) at different simulation volumes are 100.14 ± 12.920 nM (1 fl), 97.831 ± 36.831 nM (0.125 fl) and 49.639 ± 72.868 nM (0.015625 fl).

3.2.   The effect of different buffer systems

As buffers determine the autocorrelation time of the Ca^{2+} 'noise', the buffer composition of a system can possibly influence the gating of calcium dependent calcium release channels. Therefore, we here consider systems with differing buffer compositions and an IP_3R calcium channel implementing a positive calcium feedback loop. Real cellular systems comprise calcium buffers with a wide range of kinetic constants. To take this variability into account, we chose to model two 'artificial' buffers with kinetic constants that differ by a factor of 100, a fast buffer B_f $(k^+ = 0.1\,\mu M^{-1} ms^{-1})$ as well a slow buffer B_{sl} $(k^+ = 0.001\,\mu M^{-1} ms^{-1})$. Figure 3 shows sample paths of the calcium concentration and the resulting differences in the gating behavior of the IP_3 channel. Different phenomena can be observed. First, due to the fast release of relatively large amount of Ca^{2+}, the IP_3R channel induced Ca^{2+} elevation is buffered more effectively by the B_f-system than by the B_{sl}-system. The peak Ca^{2+}-concentration differs by a factor of 10. This is a substantial difference, considering the possible influence on subsequent signal pathways. Furthermore, isolated fast buffer systems limit the positive feedback of free Ca^{2+} ions on the IP_3R channel and lead to a reduced open probability of the channel. In this case, we found an open

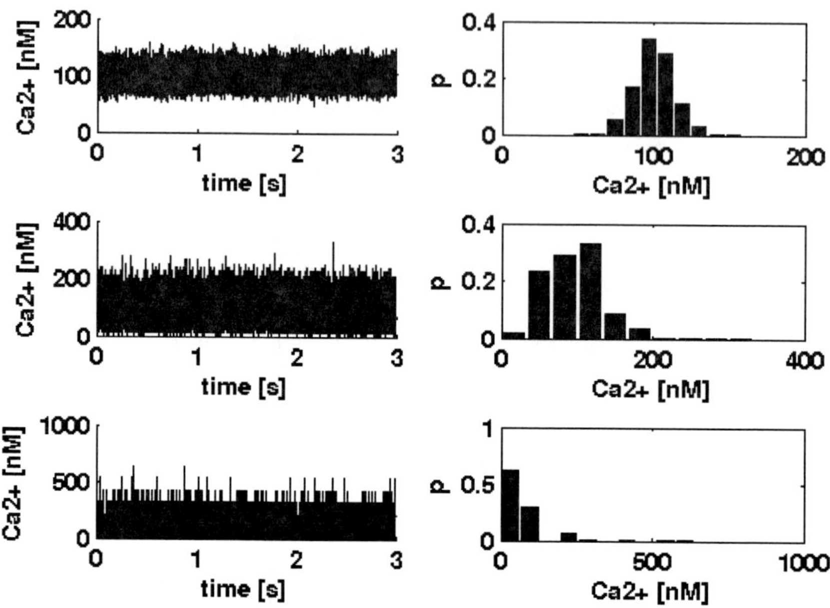

Figure 2. Effects of different simulation volumes: in these simulations, the only parameter varied was the system volume. The left column shows exemplary Ca^{2+} paths, the right column shows the corresponding concentration histogram. Upper row: $\Omega = 1$ fl, $[Ca^{2+}] = 100.14 \pm 12.92$ nM. Note that the concentration distribution is approximately Gaussian. Middle row: $\Omega = 0.125$ fl, $[Ca^{2+}] = 100.14 \pm 36.831$ nM. Lower row: $\Omega = 0.0165$ fl, $[Ca^{2+}] = 49.6391 \pm 72.868$ nM. Compared with the 1 fl simulation, the concentration distribution is highly skewed (Poissonian).

probability of $P_o = 0.41 \pm 0.13$ for the B_{sl}-system and $P_o = 0.27 \pm 0.12$ for the B_f system. The results are based on five simulation paths each.

3.3. Calmodulin Kinase IIα

In order to illustrate possible effects of local Ca^{2+} noise on connected signalling pathways, we included calmodulin (CaM) as well as the CaM dependent kinase IIα (CaMKIIα) to represent a central cellular signalling system with a special relevance in central neurons [?]. Here, calmodulin provides the link between Ca^{2+} dynamics and the activity of the kinase. Calmodulin has four Ca^{2+} binding sites and CaMKIIα activation occurs when all four CaM sites are occupied by a Ca^{2+} ion. Thus, our signal cascade contains five reactions (four Ca^{2+} association reactions and one CaMKIIα activation reaction).

Figure 4 illustrates a sample simulation, combining the Ca^{2+} concentration (upper trace), the CaMKIIα concentration (middle trace) and the IP_3R channel state (lower trace). We chose a simulation volume of $(0.5 \ \mu m)^3 = 0.125$ fl, an initial Ca^{2+}-concentration of $0.1 \ \mu M$ and a constant IP_3-concentration of $5 \ \mu M$. Note that all channel subunits are initially in the pre-activated state [110] to produce simulations with an early onset channel

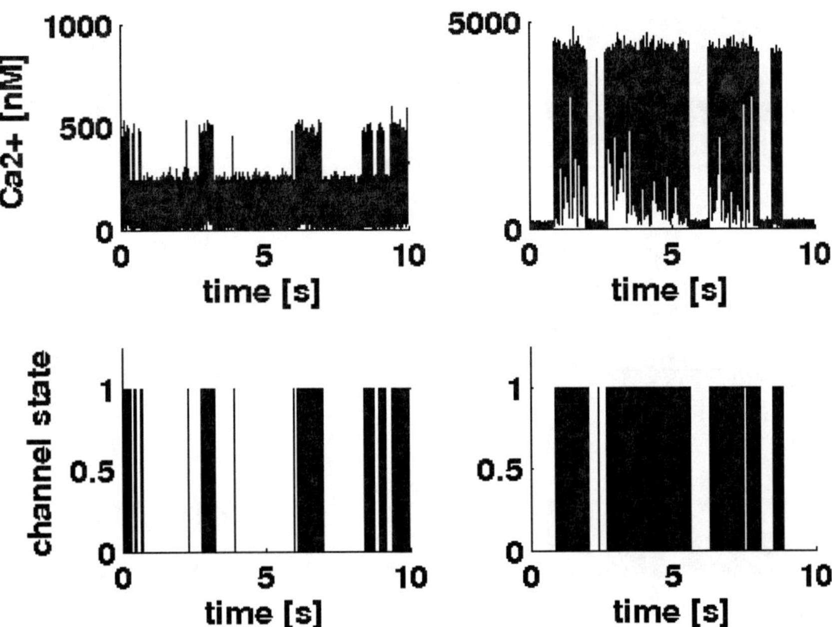

Figure 3. The effect of buffer kinetics: we introduced two artificial buffers to investigate the effects of different buffer reaction kinetics. The sample path on the left was generated in the presence of a fast buffer B_f, the right sample path in the presence of a slow buffer B_{sl} (for parameters, see text). Two phenomena are observed. The peak Ca^{2+} concentration of the two simulation paths varies by a factor of 10 and the open probability is higher for the B_{sl} system ($P_o = 0.41 \pm 0.13$) than for the B_f system ($P_o = 0.27 \pm 0.12$)

opening event. The parameter settings for these simulations are given in Tables 1 and 2. It is observed that the average number of activated CaMKIIα molecules reliably increases after activation of the calcium release channel, however, comparison of individual sample paths reveals a substantial variability in the response. It can be assumed that this variability will be reflected in the activation profiles of other, downstream signalling molecules influenced by the CaMKIIα concentration. On the other hand, our example shows that the initially highly variable Ca^{2+} signal is filtered by the CaM reaction kinetics, and is passed on as a delayed and smoothed transform of the Ca^{2+} signal. The mentioned variability is summarized in Figure 5 that shows a superposition of $n = 10$ CaMKIIα sample paths.

3.4. Computation Time

The simulation software we used for all computations is a beta-version which has been developed by our group. The most important factor influencing the computation time are the diffusion events which outnumber reaction events by a factor of at least 10^3. In addition, computation time grows nonlinearly with the system volume, especially for bimolecular reactions. The specific rate constants of the reactions considered have the lowest impact on total computation time.

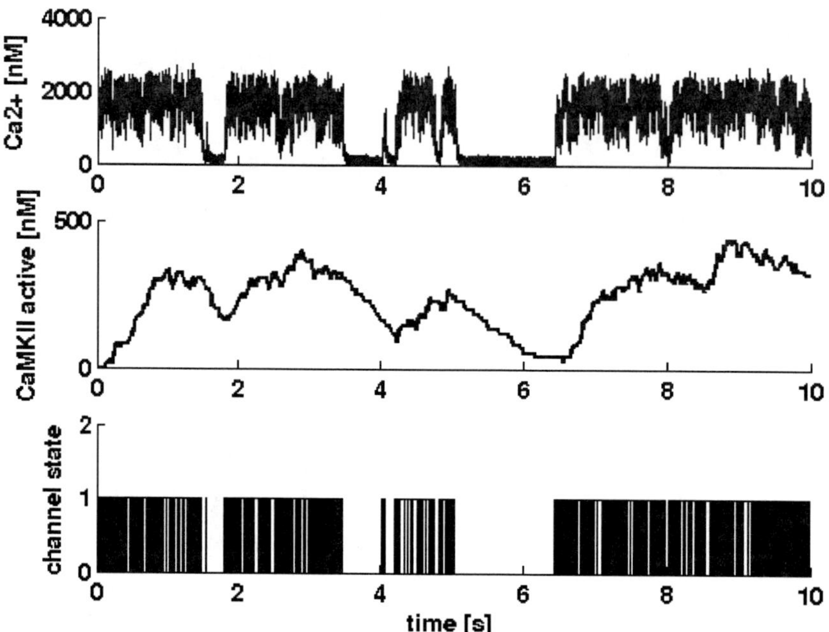

Figure 4. Simulations including a single IP$_3$R channel and the CaM-CaMKIIα pathway. The upper trace shows the Ca^{2+} concentration, the middle trace the activated CaMKIIα concentration and the lower trace the channel sate (1 open, 0 closed). Openings of the IP$_3$R calcium channel lead to a sudden increase of the local Ca^{2+} concentration. The concentration of activated CaMKIIα is a filtered version of the highly variable Ca^{2+} signal, modulated by the CaM reaction kinetics.

In section 3.1., simulation times were 1878 ± 35.7 s, $(1$ fl, $t = 3$ s$)$, 917.4 ± 5.3 s $(0.125$ fl, $t = 3$ s$)$ and 453 ± 2.7 s $(0.016525$ fl, $t = 3$ s$)$. We assume $T_s \sim \Omega$ where T_s represents the total simulation time and Ω is the simulation volume. In section 3.2., we used different buffer systems to emphasize their effects on the open probability of the IP$_3$R channel. Simulation times were 9013 ± 65.3 s, $(B_{sl}, 0.125$ fl, $t = 10$ s$)$ and 9046 ± 60.8 s $(B_f, 0.125$ fl, $t = 10$ s$)$. The most extensive simulation was presented in section 3.3.. The mean computation time was 14261 ± 109.7 s $(0.125$ fl, $t = 10$ s$)$. All simulations were carried out on a MacBook Pro with a 2.26 GHz Intel Core Duo processor and 4GB 1067 MHz DDR3 RAM. Further steps for optimization in terms of speed are currently being developed.

4. Fast Approximation of Equilibrium Fluctuations with the Chemical Langevin Equation

The Gillespie algorithm as described above generates stochastically exact sample paths of the model system. The computational cost to calculate a sample path increases with the number of possible reactions per time unit. This can be seen directly from Eq. (1)

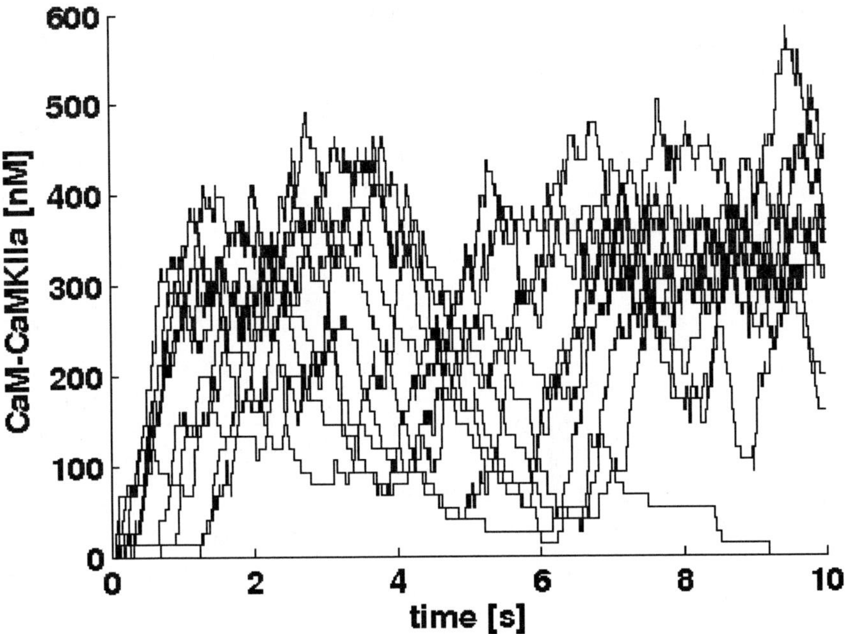

Figure 5. Stochasticity of the CaMKIIα response: the CaMKIIα activation profiles of $n =$ 10 simulations runs with identical initial conditions are shown. The sample paths are highly variable due to the combined variability of the local Ca^{2+} concentration and stochastic channel openings. As initial condition, we put all channel subunits in the state [110] to induce early onset channel openings.

which samples an exponential distribution whose mean value is $a_j(x,t)^{-1}$, i.e. the inverse of the event propensity. In the model systems discussed, the most important factors that influence the computation time are the simulation volume and the total buffer concentration. Due to the combinatorial terms that arise in the computation of reaction rates, in the case of second- and higher order reactions the computational cost grows faster than linearly with increasing simulation volume. Another decisive factor is the modeling of diffusion events. As diffusion rates often lie several orders of magnitude above buffer association and dissociation rates, the spatially resolved Gillepsie algorithm runs significantly slower than the 'classical', diffusion-free algorithm. We will now discuss the chemical Langevin equation (CLE) as a useful tool to massively increase the computation speed in the case of equilibrium Ca^{2+} fluctuations [?, ?, ?].

The simplest approximation of the local Ca^{2+} concentration is to assume the concentration to be equal to its equilibrium value and to be constant in space and time. In terms of statistical moments, this represents a first order approximation. Interpreting the instantaneous Ca^{2+} concentration as a random variable, this means that the stochastic process $[Ca^{2+}]_t$, $t \in \mathbb{R}^+$ defined by the Ca^{2+} concentration at each time t is represented by its mean value only. Second moments, i.e. the autocorrelation structure or 'noise color' however, can decisively shape the response of a system [?]. In the context of Markovian diffusion processes, the Langevin equation approach has a long history [?]. In general, a

Langevin equation is a multivariate, non-linear stochastic differential equation describing the evolution of a coarse-grained, but still stochastic version of the underlying microscopic process. While the Gillespie algorithm computes a path that is a solution of the chemical master equation, the Langevin equation is equivalent to the corresponding Fokker-Planck equation, i.e. a second-order truncation of the master equation. For numerical simulations, it is convenient to interpret the CLE as an Ito equation, a direct consequence of which is that we always obtain a multivariate Markov process [?].

The CLE describes the evolution of each component $x_i(t)$ of the state vector $\mathbf{x}(t)$ as

$$x_i(t+\mathrm{d}t) = x_i(t) + \sum_{j\in R} v_{ji}a_j(\mathbf{x}(t))\,\mathrm{d}t + \sum_{j\in R} v_{ji}\sqrt{a_j(\mathbf{x}(t))}\mathrm{d}B(t) \qquad (4)$$

where j is the index of the reaction set R, v_{ji} is the state change of molecular species i under reaction j, $a_j(\mathbf{x}(t))$ is the reaction propensity, and $\mathrm{d}B(t)$ are the increments of a standard Brownian motion. Note that the complexity of the Gillespie algorithm has been simplified in two ways. First, the exponentially distributed random waiting time τ for the next reaction event has been substituted by a fixed time increment $\mathrm{d}t$. Second, the number of reactions of type R_j that occurs in the interval $[t, t+\mathrm{d}t]$ is approximated by a random number. In the CLE framework, the approximation is given by random numbers sampled from a normal distribution with mean and variance equal to $a_j\mathbf{x}(t)\,\mathrm{d}t$. The second simplification demands $\mathrm{d}t$ to be large enough to approximate the number of reactions that occur in the interval $[t, t+\mathrm{d}t]$ by a normally distributed random number. In contrast to the Gillespie algorithm, the resulting number of reactants at a given time is no longer a natural but a real number. Multiplication with v_{ji} is necessary to implement the actually resulting change of the state vector $\mathbf{x}(t)$. For a derivation of the CLE, the reader is referred to ref. [?, ?]. Note that the change of the state vector $d\mathbf{x}(t)$ can be split into a deterministic and a stochastic or 'noise' term (first and second sum of the rhs in Eq. (4), respectively). The deterministic term is identical to the corresponding deterministic reaction rate equations of the macroscopic system.

In order to compare the results of the Gillespie algorithm and the CLE, the CLE was numerically integrated using the Euler-Maruyama method and an integration time of $\mathrm{d}t = 0.01$ ms. The results of the CLE were compared with sample paths of the same length generated by the Gillespie algorithm as described above. Parameter settings for both algorithms were as in Tables 1 and 2, the results were sampled at 100 kHz. The simulated Ca^{2+} concentration time courses of 100 ms length were quantified by their mean, their standard deviation and their autocorrelation time τ_c. The autocorrelation time was calculated from the lag-1 partial autocorrelation coefficient p_1 and the integration time interval $\mathrm{d}t$ as $\tau_c = -\frac{\mathrm{d}t}{\log(p_1)}$. The results calculated from 20 sample paths for each algorithm gave a calcium concentration of 99.92 ± 0.96 nM (SSA), 100.8 ± 5.3 nM (CLE), an average standard deviation of 12.81 nM (SSA) and 13.84 nM (CLE), and an autocorrelation time (mean $\pm$ SD) of $88.2 \pm 4.1\,\mu s$ (SSA) and $109.4 \pm 10.9\,\mu s$ (CLE). In summary, there is a very good agreement between both algorithms for the given parameter settings. In order to quantify the 'calcium noise color' by the first partial autocorrelation coefficient of the $\left[Ca^{2+}\right]$ time course, the implicit assumption of a simple first-order autoregressive process (AR(1)) is made. To justify this approximation, Fig. 6 shows exemplary Ca^{2+} time courses (left) generated by both algorithms as well as their power spectral densities (right, black solid lines) and the expected Lorentzian form of the spectrum (right, blue solid line). The results from both algorithms show an excellent fit.

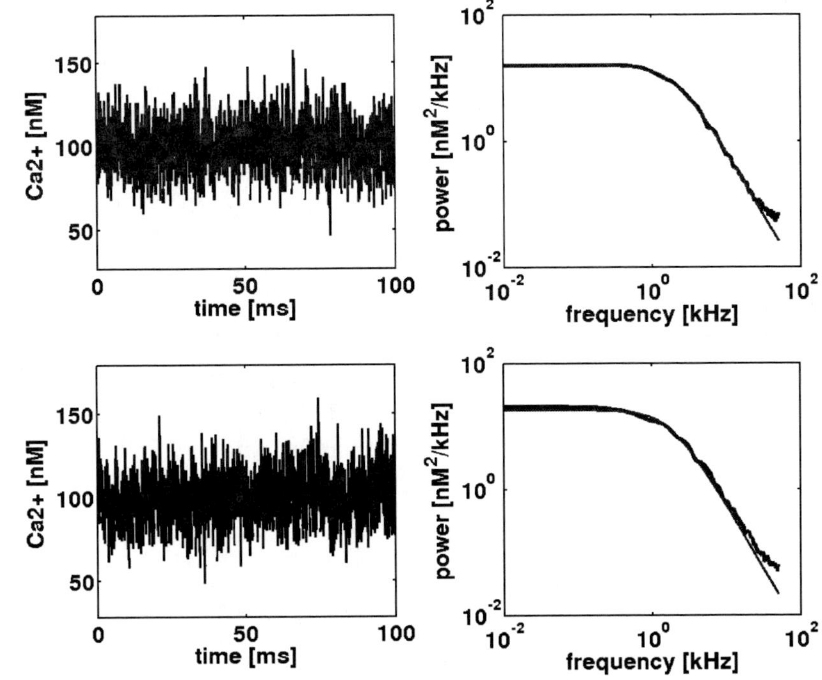

Figure 6. Comparison of the Gillespie algorithm and the chemical Langevin equation (CLE). The upper row shows the timecourse of free Ca^{2+} (left) simulated with Gillespie's algorithm and the corresponding power spectral density (PSD, right). The PSD was computed numerically (black line) and fitted to the expected Lorentzian function (blue line). The lower row shows analogous results for the CLE simulation. For the parameter settings chosen (Table 1, no channels, $\Omega = 1$ fl), both algorithms show an excellent agreement.

Why should one care about an approximate solution of the chemical master equation if an exact solution is at hand? The simple answer is speed. Simulations of subcellular signalling cascades often include the local Ca^{2+} concentration as Ca^{2+} is a key regulator of a large class of ion channels, kinases and transcription factors. However, the degree of approximation ranges from a constant Ca^{2+} assumption (non-excitable cells) to detailed stochastic simulations (muscle, neurons). When various Ca^{2+} buffers are included in simulations, the most common strategy is to use deterministic reaction rate equations and to neglect stochastic effects. However, as local signalling cascades mostly occur in sub-femtoliter volumes and the average Ca^{2+} concentration is low (ca. 60 ions/fl), stochastic effects should be considered as far as possible. To date, only few studies investigating the role of Ca^{2+} stochasticity in subcellular signalling cascades exist [?, ?, ?, ?, ?, ?]. As the SSA can increase the simulation time considerably, we propose the CLE as a fast alternative to include local Ca^{2+} fluctuations in simulations of subcellular systems. It should be noted that the treatment of diffusion events as simple reactions allows a straightforward inclusion of diffusion into the CLE framework. Note also that the CLE approach becomes invalid at very small volumes or when species with low copy numbers are included (e.g. ion channels). If the calcium dynamics are largely uncoupled from the remaining system however, the calcium-buffer subsystem can still be simulated at high speed using the CLE and be fed back into the other subsystem.

5. Conclusion

The here presented article has shown how to implement subcellular calcium dynamics, including buffering, diffusion and calcium dependent ion channels, in an exact stochastic simulation framework. We illustrated how several factors (buffer kinetics, volume) influence the simulation results and showed that the simulated paths depict a high variability based on the intrinsic stochasticity of the underlying physico-chemical processes. The variability is passed on to downstream signalling pathways (here CaM-CaMKIIα) and intermediary reactants (here CaM) act as a low-pass filter of the incoming 'calcium noise'.

We also showed that, using the chemical Langevin equation (CLE), the local calcium signal can be treated as a colored noise input to other signal pathways. The main advantage of the CLE approach is a substantial acceleration compared to Gillespie's algorithm. With these approaches, we plan to further investigate the effects and the putative biological role of local Ca^{2+} noise on more complex signalling cascades in specific cell types.

6. Appendix

Table 1. Kinetic constants and diffusion coefficients

Reactant	k^+ $(\mu Mol^{-1}ms^{-1})$	k^- (ms^{-1})	$K_D(\mu Mol)$	D $(\mu Mol^2 s^{-1})$	Reference
Ca^{2+}				200	[?]
IP_3				200	[?]
B_s	0.05	0.1	2	-	[?]
B_m	0.005	0.01	0.15	200	[?]
B_{dye}	0.15	0.3	2	15	[?]
CaM	0.0025	0.05	20	-	[?]
CaMCa	0.08825	0.05	1760	-	[?]
$CaMCa_2$	0.0125	1.25	0.01	-	[?]
$CaMCa_3$	0.25	1.25	0.2	-	[?]
CaMKIIα	0.0021	0.001	2.1	-	[?]

Table 2. Kinetic parameters of IP_3R calcium channel subunit transitions.

reaction index i	$a_i \ (\mu Mol^{-1} ms^{-1})$	$b_i \ (ms^{-1})$
0	0.55	0.08
1	0.08	0.64
2	$4*10^{-5}$	$4.8*10^{-4}$
3	0.08	40
4	$4*10^{-4}$	$7.68*10^{-5}$
5	0.015	0.012

In: Calcium Channels: Properties, Functions and Regulation
Editor: Mark R. Figgins

ISBN: 978-1-61470-232-0
©2012 Nova Science Publishers, Inc.

Chapter 6

LETHARGIC MICE: AN INTERESTING ANIMAL MODEL TO INVESTIGATE VOLTAGE-GATED CALCIUM CHANNEL β_4 SUBUNIT FUNCTION

Katell Fablet and Michel De Waard*

[1]Unité Inserm U836, Grenoble Institute of Neuroscience, Site Santé, Cedex, France
[2]Université Joseph Fourier, Grenoble, France

ABSTRACT

Lethargic (*lh*) mice are spontaneous mutant mice presenting a loss of β_4 subunit of voltage-gated calcium channels (VGCC) due to a mutation in the gene encoding for β_4, Cacnb4. These *lh* mice are characterized by a complex phenotype with notably severe neurobehavioral defects. They comprise gait ataxia, paroxysmal dyskinesia, hypokinetic behavior and absence epilepsy seizures. The epileptopathogenesis has been extensively investigated in this mutant strain. Disturbances of the $GABA_B$ receptor system with the involvement of T-type VGCC in the thalamus may explain the appearance of absence epilepsy. Considering the role of β subunits in VGCC assembly, targeting and modulation of voltage-sensitive parameters, loss of β_4 has severe consequences for VGCC functions. Data from *lh* mice gave insight into an interesting phenomenom called "subunit reshuffling". Indeed, a compensation mechanism by β subunits has been observed in certain brain regions of *lh* mice. Far from being complete, "subunit reshuffling" is evidence of neuronal plasticity that may attenuate the severe defects caused by the absence of β_4 in *lh* mice. Where compensation is not possible, the absence of β_4 highlights additional roles of β_4 possibly in gene regulation.

---

* Correspondence should be addressed to Katell Fablet, E-mail : katell.fablet@ujf-grenoble.fr, Tel.: +0033 456520567; fax : +0033 456520669

INTRODUCTION

Voltage-gated calcium channels (VGCC) play many biological functions, such as hormones and neurotransmitters release, neuromuscular coupling and gene regulation. These channels comprise alpha1 (α1 or Ca$_v$), a large membrane subunit constituting the pore-forming subunit of VGCC and up to three auxiliary subunits of type beta (β), gamma (γ), and alpha2delta ($\alpha_2\delta$) (Fig. 1A). Spontaneous mutant mice strains constitute tremendous animal models to study the organization and the role of VGCC. The *tottering, learner, rolling* or *Nagoya rocker* mutant mice were identified for a mutation in the Cacna1 gene encoding the α1 subunit of P/Q-type VGCC (Fletcher, Lutz et al. 1996; Mori, Wakamori et al. 2000). In addition, *stargazer* and *lethargic* (*lh*) mice have been respectively described for a γ_2 and a β_4 mutation (Burgess, Jones et al. 1997; Letts, Felix et al. 1998). This chapter will focus on *lh* mice and aim at describing the neuropathological defects affecting this strain mutant mice and delineating neural networks and cellular mechanisms at the origin of the *lh* phenotype. In addition, this review will evidence that β_4 role is not limited to assembly and regulation of VGCC.

LETHARGIC MICE DISPLAY A COMPLEX PHENOTYPE

First described by Dickie in 1964, *lh* mice, which spontaneously arose from the BALB/cGn mice line, show a complex neurobehavioral phenotype. These mice exhibit a difficulty in moving forward with a slow and hesitant motor behavior (Dickie 1964). It is because of this behavior that the name *lh* has been given to this mice strain. These mice were early characterized by the spontaneous seizures that they exhibit (Dickie 1964). Electroencephalographic studies revealed bilateral synchronous electrographic seizures of 5- to 6-Hz spike waves. Epileptiform bursts were considered as spike-wave discharges (SWD) if they met all the following criteria : duration of bursts >0.6 sec with epileptiform spikes (i.e., <70 msec per spike), a typical frequency for mice of 5-6 Hz and simultaneously recorded in the both hemispheres (Hosford, Lin et al. 1995). For the duration of seizures, *lh* mice stay immobile and unresponsive. In humans, typical absence seizures are characterized by brief episodes of bilaterally synchronous 2.5 to 4 Hz spike wave discharges that manifest themselves as sudden behavioral arrest and impaired consciousness (Weiergräber, Stephani et al. 2010). These characteristics resemble those observed in *lh* mice and because of this resemblance, this mutant strain is considered as a valid and useful model to study absence epileptic seizures. The hypokinetic behavior of the mice turns out more to be a strategy learned by the mice to avoid attacks than the result of an impaired neuromotor system (Khan and Jinnah 2002). This hypothesis is supported by the fact that mice display fewer seizures over 1-3 months of age compared to younger mice, a time lapse coincident with the emergence of this hypokinetic behavior. Other neurologic defects have been described for this strain mutant as well. *Lh* mice exhibit an ataxic gait at 2-3 weeks (Sidman, Green et al. 1965). Moreover, it has been shown that *lh* mice display paroxysmal dyskinesia characterized by intermittent attacks of involuntary abnormal movements superimposed on a relatively normal baseline (Devanagondi, Egami et al. 2007; Shirley, Rao et al. 2008). This defect primarily confused with manifestations of motor epilepsy has been clearly distinguished from absence epileptic seizures. Briefly, the distinguishable characteristics that were identified are the EEG records not reflecting epilepsy, the duration of seizures (longer than epileptic seizures), the absence of presumed unconsciousness accompanying absence epileptic seizures, the possibility of repeated seizures without the usual postical refractory period detected in motor epilepsy and, finally, the trigger of attacks by environmental factors not observed for epileptic seizures.

To that complex panel of neurologic disorders, these mice suffer also from other defects regarding body growth, and the immune and reproductive systems. Indeed, mice show a postnatal growth retardation with a reduction in body weight (Dung and Swigart 1972; Dung 1975) and cachexia (i.e. syndrome comprising a loss of weight, muscle atrophy, and weakness) between 3 and 6 weeks of age (Dung and Swigart 1972). The immune system is largely affected in this mutant strain as well (Dung and Swigart 1972) with, among other symptoms, an abnormal thymus development, a smaller spleen and decreased lymphocyte cell numbers. Finally, these mice also suffer from low-fertility. Maintenance of the strain is therefore difficult. The generation of homozygote *lh/lh* is possible almost exclusively by crossing *lh*/wild-type (WT) heterozygotes. Because of the sum of all these defects, *lh* mice are hit by premature death with a prevalence of high mortality rates between 3-4 weeks of age. It is nevertheless important to notice that *lh* mice, surviving past two months of age, recover much of their body weight and immune functions, suggesting that these defects are age-related.

A MUTATION IN THE β_4 SUBUNIT OF VOLTAGE-GATED CALCIUM CHANNEL GENE IS AT THE ORIGIN OF THE *LH* PHENOTYPE

Burgess et al. were the first in 1997 to highlight the molecular defect underlying the *lh* phenotype. This research group identified a mutation within the gene of β_4 subunit of VGCC, Cacnb4 on chromosome 2 (Burgess, Jones et al. 1997). β subunits are encoded by four distinct genes. Thus four isoforms of β subunits, including β_4, have been identified (Fig. 1B). Each one of these isoforms also contains several splice variants (Buraei and Yang 2010). β subunits have a modular structure and can be divided in five domains. Domains D2 and D4 are particularly conserved among β isoforms and correspond respectively to the Scr Homology 3 (SH3) and Guanylate Kinase (GK) domains which are linked by a HOOK sequence. These structural features of β subunits suggest that these subunits belong to the larger family of membrane-associated guanylate kinases (MAGUK). β_4 was first cloned by Castellano et al. in 1993 (Castellano, Wei et al. 1993) and his crystal structure has been found in 2004 by Chen et al. (Chen, Li et al. 2004, Fig. 1B). This protein presents on Western blot a molecular weight of approximately 58 kD. β subunits chaperone the pore-forming subunit to the cell membrane for VGCC assembly and targeting and sets voltage-sensitive parameters for channel opening and closing (De Waard and Campbell 1995; Brice, Berrow et al. 1997). The *lh* mutation corresponds more precisely to the insertion of four nucleotides within a 5' splice site of Cacnb4 gene (Fig. 1C). This mutation leads to a translational frame shift and 77 bp exon preceding the mutation is consistently skipped during Cacnb4 pre-mRNA processing in *lh* brain. This predicts a truncated β_4 subunit of VGCC protein missing 60% of the C-terminal sequence relative to the wild-type protein. However to date, attempts to show such a truncated sequence expressed in neurons were unsuccessful. McEnery et al. failed to detect by Western blotting this truncated β4 subunit in forebrain and cerebellum of *lh* mice (McEnery, Copeland et al. 1998). The absence of this truncated sequence indicates that the translated product is highly unstable and degraded rapidly after synthesis. *Lh* mice is therefore considered by many as a spontaneous β4 knock-out model. This assumption should however be carefully accepted considering that if such a truncated β4 sequence would be expressed, it most likely could possess some functional activity as much as some short splice variants of the β4 subunit, such as β4c (Hibino, Pironkova et al. 2003; Xu, Lee et al. 2011).

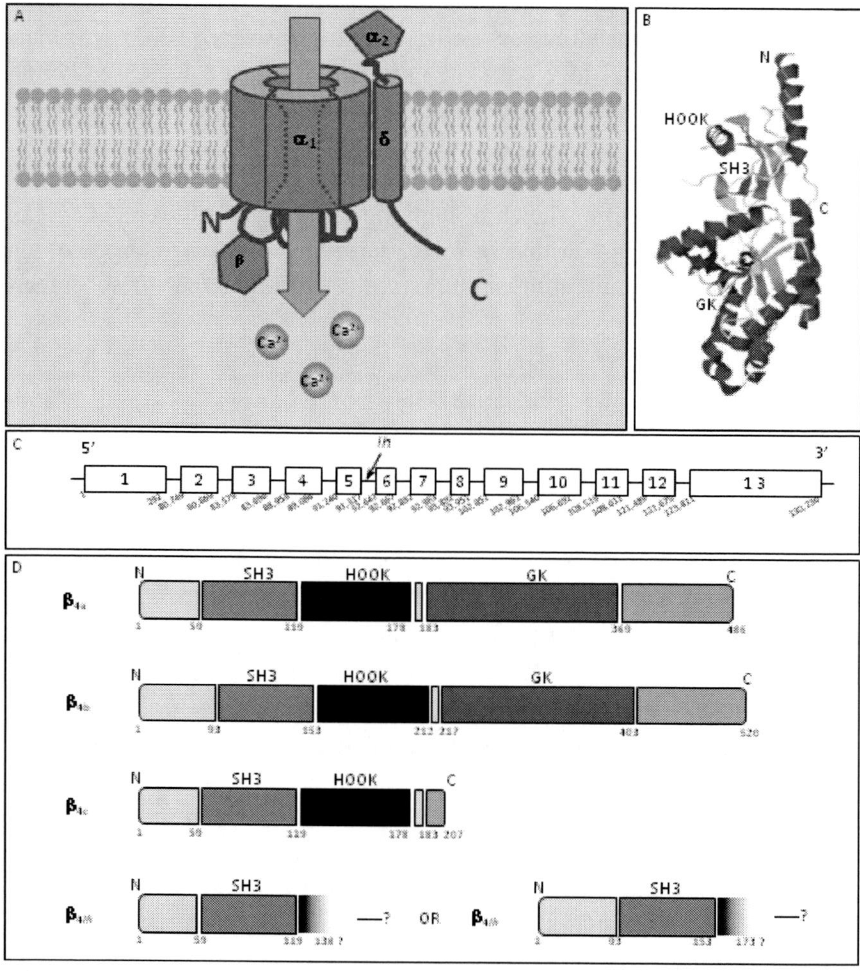

Figure 1. Auxiliary subunits making up VGCC and the different β splice variants. A- VGCCs comprise α1, a large membrane subunit and up to three auxiliary subunits of type β, γ, and α2δ. B- Cristal structure of β4 from Chen et al. (Chen, Li et al. 2004). C- Schematic DNA sequence organization of Cacnb4 with exons (white rectangles), the position of the lh mutation is represented by an arrow. D- β4 subunits form part of MAGUK protein family and contain notably SH3 and GK domains. Three β4 splice variants have been identified in human, β4a, β4b, β4c. Hypothetical structure of the truncated β4 from lh mice is also represented. The C-terminal sequence of this putative protein is yet inknown

LH MICE: A MODEL OF ABSENCE EPILEPTIC SEIZURES

Considered as an interesting mouse model of absence epilepsy, *lh* mice were studied to delineate neural networks and molecular and cellular mechanisms at the origin of this phenotype. In addition, *lh* mice were also used as an animal model to develop new antiepileptic drugs. Early, the implication of GABA$_B$ receptors in the genesis of seizures has been highlighted. Indeed the administration of GABA$_B$ receptor agonists and antagonists respectively increases and decreases seizures frequency in *lh* mice (Hosford, Clark et al. 1992; Hosford, Lin et al. 1995; Hosford, Wang et al. 1995; Aizawa, Ito et al. 1997). It has been shown that GABA$_B$ receptor density increases in the neocortex but also in multiple nuclei of thalami and in caudate-putamen of *lh* mice compared to WT mice (Hosford, Clark et al. 1992; Hosford, Lin et al. 1995; Hosford, Lin et al. 1999). This is not the case for GABA$_A$ receptors whose numbers remain unchanged in the neocortex of *lh* mice. In the same way, significantly higher GAD$_{67}$ (a glutamate decarboxylase that synthesizes GABA from glutamate) mRNA and protein expressions have been detected in the nucleus reticularis (NRT) of the thalamus of *lh* mice (Lin, Lin et al. 1999). These data suggest that there is an increase in GABA synthesis in NRT but unfortunately this question has not been investigated in detail by complementary studies. There is however a study that explored amino acid levels in different brain areas of *lh* and WT mice (De Luca, Di Giorgio et al. 2005). The concentration of GABA in the diencephalon remains unchanged between *lh* and WT mice but there is a significantly higher concentration of GABA in the cortex and a significantly lower concentration in the cerebellum of *lh* mice. Modifications concerning GABA concentration and increase of GABA$_B$ receptors are not correlated with clear alterations in GABA release or GABA$_B$ receptor mediated effects. Hosford et al. explored in 1992 the possibility that GABA$_B$ receptor mediated effects contribute to the neuronal mechanisms underlying absence seizures. They observed a significantly greater inhibition of N-methyl D-aspartate (NMDA) excitatory postsynaptic potentials (EPSPs) by GABA$_B$ receptor in CA1 of hippocampal slice of *lh* brain (Hosford, Clark et al. 1992). In neocortex of *lh* mice, presynaptic GABA$_B$ receptors inhibit to a significantly greater degree [^{3}H]-GABA release (Lin, Wang et al. 1995). On the contrary in the thalamus of *lh* mice, the presynaptic GABA$_B$ receptor-mediated effect is reduced. In thalamic ventrobasal neurons, whole cell voltage-clamp recordings showed no differences in evoked GABA$_B$-mediated currents in *lh* or wild-type mice (Caddick and Hosford 1996). Similarly, Caddick et al. observed in 1999 no differences in effects mediated by presynaptic GABA$_B$ receptors on inhibitory afferents to ventrobasal neurons in *lh* mice (Caddick, Wang et al. 1999). Even if GABA receptor system defects seem complex, alterations of this system in the thalamus of *lh* mice might be predominantly involved in the pathogenesis of absence epilepsy. It is now well established that the thalamus contributes to epileptogenesis (Kohl, Paulsen et al. 2010). Reciprocally, corticothalamic projections permit synchronized network activity that explains synchronous pathological spike-and-wave discharges during generalized absence epilepsy in the cortex and in the thalamus. Candidate structures regulating absence seizures in *lh* mice have been identified in the thalamus: the antero-ventro-lateral nucleus, the reuniens nuclei and the reticular nucleus (Hosford, Lin et al. 1995). These structures present shared features: they contain enriched GABA$_B$ binding sites, present spike-wave discharges in frontal neocortex, synchronous with spike-wave discharges in thalamic nuclei, and show regulation of absence seizures by GABA$_B$ receptors. The involvement of reticular nucleus in the genesis of spike and wave discharges in absence epilepsy has been demonstrated in other animal models and particularly in the Genetic Absence Epilepsy Rats from Strasbourg (GAERS) (Tsakiridou, Bertollini et al. 1995; Slaght, Leresche et al. 2002). The reticularis nucleus is exclusively composed of GABAergic neurons that project to the dorsal thalamus and receive axon collaterals from both corticothalamic and thalamocortical fibers. GABAergic neurons of the reticularis nucleus exert influences on thalamocortical neurons and local-circuit GABAergic cells and more precisely hyperpolarize these cells (Steriade and Contreras 1995). This hyperpolarization is a prerequisite for de-inactivation of T-type calcium channels thereby enhancing the generation of rebound action potential burst (Kohl, Paulsen et al. 2010). The implication of T-type calcium currents in the generation of spike and wave discharges characteristic of absence epilepsy has already been reported in GAERS (Danober, Deransart et al. 1998; Talley, Solórzano et al. 2000). In *lh* mice, recordings of thalamocortical neurons showed higher T-type Ca^{2+} current levels compared with WT mice (Zhang, Mori et al. 2002). This enhancement of T-type currents is not explained by a variation in $Ca_v3.1$, $Ca_v3.2$ or $Ca_v3.3$ mRNA expression. Another study investigated the role of $Ca_v3.1$ in the genesis of absence epilepsy in *lh* mice. For this purpose Song et al. constructed a double mutant $Ca_v3.1^{-/-}/\beta_4^{lh/lh}$ mice (Song, Kim et al. 2004). They showed that the $Ca_v3.1$ null mutation drastically reduces spike and wave discharges in *lh* mice revealing the important role of $Ca_v3.1$ in the genesis of absence seizures in *lh* mice.

RESHUFFLING OF β SUBUNITS COMPENSATE THE LACK OF β₄ IN *LH* MICE

β_4 seems to preferentially associate with $Ca_v2.1$ and $Ca_v2.2$ of respectively, P/Q- and N-type channels (Liu, De Waard et al. 1996). β_4 has been proved to interact with high affinity with the alpha interaction domain of the cytoplamic loop I-II of $Ca_v2.1$ subunit (De Waard and Campbell 1995). Moreover several studies proved a co-localization of β_4 and $Ca_v2.1$ in cerebellar Purkinje cells and granule cells (Lin, Wang et al. 1995; Ludwig, Flockerzi et al. 1997; Volsen, Day et al. 1997; Lin, Barun et al. 1999). Concerning N-type channels, association of β_4 with $Ca_v2.2$ has been observed in the forebrain and the cerebellum (McEnery, Copeland et al. 1998). In *lh* mice, the loss of β_4 would affect the preferential $Ca_v2.1$ and $Ca_v2.2/\beta_4$ interaction. This modification may itself trigger defects in P/Q- and N-type channel localization and decrease in calcium current amplitudes. Several studies investigated this question and surprisingly invalidated this *a priori* seducing hypothesis. Immunohistochemical experiments on adult cerebellum sections analyzed potential changes in the localization of $Ca_v2.1$ and $Ca_v2.2$ in WT and *lh* mice (Burgess, Biddlecome et al. 1999). In fact, Burgess et al. observed no differences between *lh* and WT mice with regard to the localization of these two pore-forming subunits. $Ca_v2.1$ immunoreactivity is mostly abundant in soma and proximal dendritic segments of Purkinje neurons and at lower levels in the molecular and granular layers. The $Ca_v2.2$ immunoreactivity is very distinct from $Ca_v2.1$. This channel type is not detected in Purkinje neurons but is present in molecular and granular layers (Burgess, Biddlecome et al. 1999). Results from *in situ* hybridization histochemistry show also an identical distribution of $Ca_v2.1$ and $Ca_v2.2$ mRNA in *lh* and WT mice brain sections (Lin, Barun et al. 1999). More precisely, $Ca_v2.1$ mRNA is highly expressed in cerebellar and dentate granule cells populations and $Ca_v2.2$ mRNA is mostly present in granule cells olfactory bulb and in the granule cells layer of cerebellum. In addition, the functions of both P/Q and N-type Ca^{2+} channels do not seem to be altered in *lh* mice. Indeed Ba^{2+} currents through VGCC in *lh* cerebellar Purkinje cells are not different from normal mice in terms of voltage-dependence of activation, voltage-dependence of inactivation and current amplitude (Burgess, Biddlecome et al. 1999). Moreover, in the hippocampal CA3-CA1 synapse, the Ca^{2+} influx profile is not significantly changed for *lh* mice (Qian and Noebels 2000). Compensatory molecular mechanisms could explain the conserved localization and biophysical properties of P/Q- and N-type channels in *lh* mice. McEnery et al. studied β subunit expression levels in forebrain and cerebellum of *lh* and normal mice (McEnery, Copeland et al. 1998). Their data indicated an increase in $β_{1b}$ expression in the forebrain and in the cerebellum of *lh* mice supporting enhancement of $Cav2.2/β_{1b}$ reassembly in *lh* forebrain. Burgess et al. showed also the presence of a subunit reshuffling and found more $β_{1b}$ and $β_3$ associated with $Ca_v2.1$ and more $β_{1b}$, $β_2$ and $β_3$ associated with $Ca_v2.2$ in *lh* mice brain (Burgess, Biddlecome et al. 1999). These modifications are not accompanied by significant changes in transcriptional regulation of the $β_{1-3}$ subunit genes. This may suggest enhanced stability of β subunits due to the association of β subunits with the pore-forming subunit or an increase in pre-existing mRNA translation. These two studies suggested that the lack of $β_4$ is compensated by the reorganization in β subunit composition of P/Q- and N-type calcium channels. This "subunit reshuffling" is in fact quite imperfect. Indeed, in spite of the reassembly of $Ca_v2.2$ with $β_{1b}$ in the cerebellum, significant decreases in N-type VGCC density and in $Ca_v2.2$ expression have been observed (McEnery, Copeland et al. 1998). This type of compensatory molecular mechanism is probably not possible in the thalamus, a brain region characterized by a strong expression of $β_4$ and by the absence of expression of the majority of other β subunits (Buraei and Yang 2010). This observation may explain why this brain area is particularly affected by the *lh* mutation at the basis of the severe neurologic defects and more particularly the absence epilepsy.

LH MICE: NEW PERSPECTIVES

Lh mice have been extensively studied as a model of absence epileptic seizures but have been underestimated as a potent model for others pathologies. Indeed, this mutant mice strain has been poorly investigated to delineate the neuroanatomical basis for paroxysmal dyskinesia, a neurologic defect clearly identified in *lh* mice (Khan and Jinnah 2002; Shirley, Rao et al. 2008). To date only one study has set out to understand neuroanatomical substrates at the origin of this neurobehavioral phenotype. This study pointed out that cerebellectomy in *lh* mice remove paroxysmal dyskinesia revealing the involvement of the cerebellum in the occurrence of this neurological disorder (Devanagondi, Egami et al. 2007). In addition, *lh* mice suffer from gait ataxia, this could be interesting to examine this coordination movement dysfunction through this mutant mice strain. Moreover, defects relative to the immune and reproductive systems, which *lh* mice suffer from, could also be interestingly explored in this mutant strain. As we have already mentioned, the effect of the mutation of $β_4$ subunit gene, which leads to the loss of $β_4$ protein, is compensated by other β subunits and notably $β_{1b}$. In brain regions, such as in the thalamus, where we can postulate that there is no compensation, targeting of VGCC to the membrane, VGCC assembly and voltage-sensitive parameters setting for channel opening and closing may be defective, implying disturbances of the calcium homeostasis. The β subunit reshuffling hypothesis, which is supported by various research groups in the world, implies that β subunits isoforms are relatively comparable in terms of plasma membrane targeting, assembly of VGCC, gating of VGCC pore-forming subunits, protein environment and cell function. However, we should wonder whether it is judicious to limit β subunits in their unique role as VGCC chaperones. We will discuss hereunder hypothetical other roles of β subunits. Indeed, several β_4 isoforms resulting of alternative splicing have been identified (Fig. 1D), and some of them have been shown to regulate gene expression. In 2002, Helton and Horne revealed that alternative splicing of the N-terminus of the β_4 subunit yields both β_{4a} and β_{4b} subunits. The structural difference between these two splice variants lies in domain D1 that contains, respectively, 15 and 49 amino acids. D1 of β_{4a} and β_{4b} could definitively not be considered as homologous (Helton and Horne 2002). β_{4a} is expressed in the spinal cord, in the medulla, the putamen, and the temporal and occipital lobes, whereas β_{4b} is expressed in the frontal lobe (Helton, Kojetin et al. 2002). Both splicing variants are expressed in the cerebellum but with different cellular distribution. β_{4a} is found in Purkinje cell dendrites and in the granule cell layer and β_{4b} is detected in Purkinje cell bodies and throughout the fibers of Bergmann glia traversing the molecular layer. A novel splice variant, a truncated β_{4C} protein of about 23 kDa has recently been identified in human brain (Xu, Lee et al. 2011). The sequence of this splice variant appears homologous to the chicken β_{4C} described in an earlier report (Hibino, Pironkova et al. 2003) and contains the β_{4a} N-terminus, the SH3 domain, the HOOK sequence, a truncated GK domain and an additional small C-terminal sequence of 13 amino acids. Chicken β_{4C} was found to regulate gene expression (Hibino, Pironkova et al. 2003). Interestingly, Western blot analysis of cerebellar/brainstem homogenates revealed the presence of the human β_{4C} in the nuclei purified fraction and immunohistochemistry experiments demonstrated the localization of human β_{4C} in vestibular nuclei (Xu, Lee et al. 2011). The same study illustrated the interaction of β_{4C} with the chromoshadow domain of heterochromatin protein 1γ (HP1γ). HP1γ is involved in gene silencing. The sequence and structure of this protein can be divided into three regions, the chromodomain, a module at the amino terminus responsible for HP1γ binding to di- and trimethylated lysine 9 of histone H3, the carboxy-terminal chromoshadow domain involved in homo- and/or heterodimerization and interaction with other proteins and a variable linker between the chromodomain and the chromoshadow, containing a nuclear localization sequence. The interaction of human β_{4C} with HP1γ may influence gene regulation. β_{4C} may indeed attenuate or on the contrary promote gene repression role of HP1γ on specific genes. This hypothesis is credible, indeed chicken β_{4C} have been shown to regulate transcriptional repression activity of chromobox protein 2/HP1γ (Hibino, Pironkova et al. 2003). Recently, Subramanyam et al. have reported the nuclear localization of β_{4b} in cerebellar granule and Purkinje cells, that strengthens a postulate role of β_4 in gene regulation (Subramanyam, Obermair et al. 2009). This observation may be related to microarray data set GSE6275 from Gene Expression Omnibus comparing gene expression in WT and *lh* mice cerebellum. The statistical analyses showed 66 genes up-regulated and 48 genes down-regulated in *lh* cerebellum (for details, see Fig. 2). Thus a new role of β_4 in gene regulation may be considered. Finally, an additional study highlighted novel roles of β_4 that should be independent of its VGCC chaperone function. This study used a zebrafish model. For this specie, two β_4 genes have been identified, $\beta_{4.1}$ and $\beta_{4.2}$ (Ebert, McAnelly et al. 2008). $\beta_{4.1}$ and $\beta_{4.2}$ morpholino knockdown have been constructed and showed an abnormal embryogenesis with defects in epiboly. This phenotype is rescued by the injection of human β_{4a} or β_{4b}, proving that human β_4 subunits are functionally efficient in zebrafish and suggesting that the cell biological functions of β_4 found in zebrafish could be relevant to other species. The functions of $\beta_{4.1}$ and $\beta_{4.2}$ in epiboly have been proved to be independent of Ca^{2+} channel activity. Thus, the complex neurobehavioral phenotype observed in *lh* mice may well be attributed to a defective role of β_4 as VGCC chaperone, but could also be linked to a default in gene regulation. It is likely that the *lh* mice will turn out useful for the investigation of these additional functional roles of the β_4 subunit.

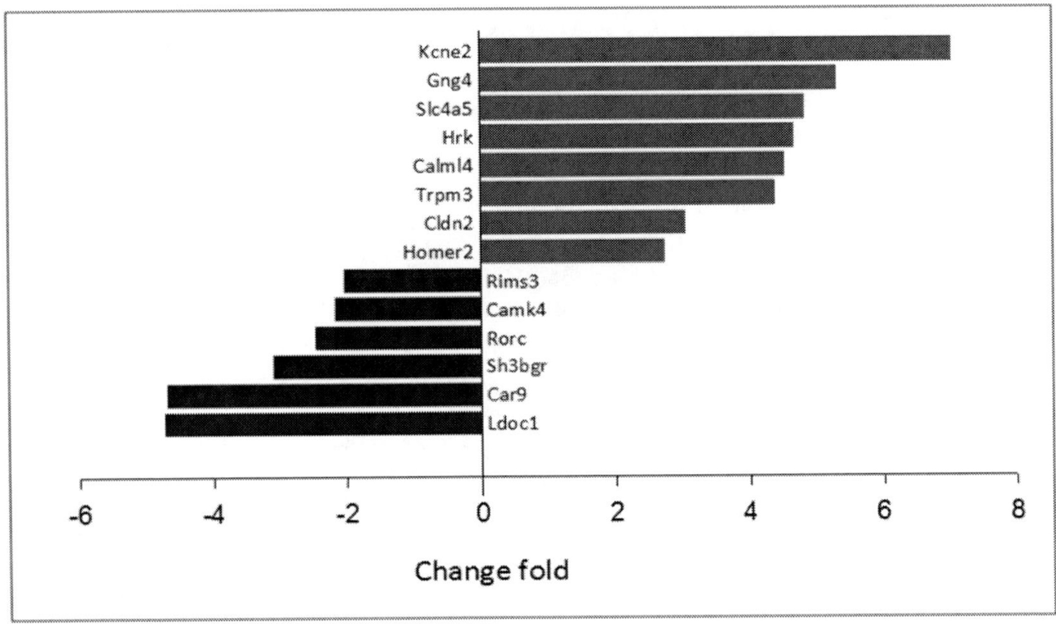

Figure 2. Differential gene expression in cerebellum of lh versus WT mice. Microarray data set GSE6275 from Gene Expression Omnibus, http://www.ncbi.nlm.nih.gov/geo/ statistically analyzed compare gene expression in WT and lh mice cerebellum. 65 genes are up-regulated and 48 genes are down-regulated regulated in lh cerebellum compared with WT (over twofold expression change, $p<0.05$). Among these genes, the potassium voltage-gated channel,Isk-related family, member 2 (Kcne2), the guanine nucleotide binding protein gamma 4 (Gng4), the solute carrier family 4, sodium bicarbonate cotransporter member 5 (Slc4a5), the harakiri (Hrk), calmodulin-like 4 (Calml4), the transient receptor potential cation channel, subfamily M, member 3 (Trpm3), the claudin 2 (Cldn2), the homer homolog 2 (Homer2) are up-regulated. In contrast the regulating synaptic membrane exocytosis 3 (Rims3), the calcium/calmodulin-dependent protein kinase IV (Camk4), the RAR-related orphan receptor gamma (Rorc), the SH3 domain binding glutamic acid-rich protein (Sh3bgr5), the carbonic anhydrase 9 (Car9), the leucine zipper doxn-regulated in cancer 1 (Ldoc) are down-regulated in lh cerebellum compared with WT.

CONCLUSION

The data from *lh* mice gave us an interesting insight into cellular mechanisms responsible for the appearance of absence epilepsy and, to a lesser extent, of paroxysmal dyskinesia. Up to now, β subunits have been mostly considered exclusively as VGCC chaperons and studies investigating *lh* mice designed in that sense have by a majority explored defects concerning VGCC. Recent studies have suggested other roles played by β subunits. The potential involvement of β subunits and particularly $β_4$ in gene regulation open new lines of research. It will be interesting to investigate the likely modification of gene regulation in *lh* mice and reconsider if this potential alteration could in part explain the appearance of the severe neurobehavioral defects which *lh* mice suffer from.

REFERENCES

Aizawa, M., Y. Ito, et al. (1997). "Pharmacological profiles of generalized absence seizures in lethargic, stargazer and [gamma]-hydroxybutyrate-treated model mice." *Neuroscience Research* 29(1): 17-25.

Brice, N., N. Berrow, et al. (1997). "Importance of the different beta subunits in the membrane expression of the alpha1A and alpha2 calcium channel subunits: studies using a depolarization-sensitive alpha1A antibody." *Eur J Neurosci.* 9(4): 749-59.

Buraei, Z. and J. Yang (2010). "The beta Subunit of Voltage-Gated Ca2+ Channels." *Physiological Reviews* 90(4): 1461-1506.

Burgess, D. L., G. H. Biddlecome, et al. (1999). "beta Subunit Reshuffling Modifies N- and P/Q-Type Ca2+Channel Subunit Compositions in Lethargic Mouse Brain." *Molecular and Cellular Neuroscience* 13(4): 293-311.

Burgess, D. L., J. M. Jones, et al. (1997). "Mutation of the Ca2+ channel beta subunit gene Cchb4 is associated with ataxia and seizures in the lethargic (lh) mouse." *Cell* 88(3): 385-92.

Caddick, S. J. and D. A. Hosford (1996). "GABAB-activated gK+ in thalamic neurons in the lethargic (lh/lh) mouse model of generalized absence seizures." *Neuroscience Letters* 205(1): 29-32.

Caddick, S. J., C. Wang, et al. (1999). "Excitatory But Not Inhibitory Synaptic Transmission Is Reduced in Lethargic (Cacnb4lh) and Tottering (Cacna1atg) Mouse Thalami." *J Neurophysiol* 81(5): 2066-2074.

Castellano, A., X. Wei, et al. (1993). "Cloning and expression of a neuronal calcium channel beta subunit." *Journal of Biological Chemistry* 268(17): 12359-12366.

Chen, Y.-h., M.-h. Li, et al. (2004). "Structural basis of the [alpha]1-[beta] subunit interaction of voltage-gated Ca2+ channels." *Nature* 429(6992): 675-680.

Danober, L., C. Deransart, et al. (1998). "Pathophysiological mechanisms of genetic absence epilepsy in the rat." *Progress in Neurobiology* 55(1): 27-57.

De Luca, G., R. M. Di Giorgio, et al. (2005). "Amino acid levels in some lethargic mouse brain areas before and after pentylenetetrazole kindling." *Pharmacology Biochemistry and Behavior* 81(1): 47-53.

De Waard, M. and K. P. Campbell (1995). "Subunit regulation of the neuronal alpha 1A Ca2+ channel expressed in Xenopus oocytes." *The Journal of Physiology* 485(Pt 3): 619-634.

Devanagondi, R., K. Egami, et al. (2007). "Neuroanatomical substrates for paroxysmal dyskinesia in lethargic mice." *Neurobiology of Disease* 27(3): 249-257.

Dickie, M. (1964). " Lethargic (lh)." *Mouse News Lett* 30: 31.

Dung, H. (1975). " Growth retardation, high mortality, and low reproductivity of neurological mutant mice. ." *Anat Rec* 181: 347-348.

Dung, H. C. and R. H. Swigart (1972). "Histo-pathologic observations of the nervous and lymphoid tissues of lethargic mutant mice." *Tex Rep Biol Med* 30(1): 23-39.

Ebert, A. M., C. A. McAnelly, et al. (2008). "Genomic organization, expression, and phylogenetic analysis of Ca2+ channel {beta}4 genes in 13 vertebrate species." *Physiol. Genomics* 35(2): 133-144.

Fletcher, C. F., C. M. Lutz, et al. (1996). "Absence Epilepsy in Tottering Mutant Mice Is Associated with Calcium Channel Defects." *Cell* 87(4): 607-617.

Helton, T. D. and W. A. Horne (2002). "Alternative Splicing of the beta 4 Subunit Has alpha 1 Subunit Subtype-Specific Effects on Ca2+ Channel Gating." *J. Neurosci.* 22(5): 1573-1582.

Helton, T. D., D. J. Kojetin, et al. (2002). "Alternative Splicing of a beta 4 Subunit Proline-Rich Motif Regulates Voltage-Dependent Gating and Toxin Block of Cav2.1 Ca2+ Channels." *J. Neurosci.* 22(21): 9331-9339.

Hibino, H., R. Pironkova, et al. (2003). "Direct interaction with a nuclear protein and regulation of gene silencing by a variant of the Ca2+-channel beta4 subunit." *Proceedings of the National Academy of Sciences of the United States of America* 100(1): 307-312.

Hibino, H., R. Pironkova, et al. (2003). "Direct interaction with a nuclear protein and regulation of gene silencing by a variant of the Ca2+-channel beta4 subunit." *Proceedings of the National Academy of Sciences of the United States of America* 100(1): 307-312.

Hosford, D., S. Clark, et al. (1992). "The role of GABAB receptor activation in absence seizures of lethargic (lh/lh) mice." *Science* 257(5068): 398-401

Hosford, D., F. Lin, et al. (1999). "Studies of the lethargic (lh/lh) mouse model of absence seizures: regulatory mechanisms and identification of the lh gene." *Adv Neurol.* 79: 239-52.

Hosford, D. A., F. H. Lin, et al. (1995). "Neural network of structures in which GABAB receptors regulate absence seizures in the lethargic (lh/lh) mouse model." *The Journal of Neuroscience* 15(11): 7367-7376.

Hosford, D. A., Y. Wang, et al. (1995). "Characterization of the antiabsence effects of SCH 50911, a GABA-B receptor antagonist, in the lethargic mouse, gamma-hydroxybutyrate, and pentylenetetrazole models." *Journal of Pharmacology and Experimental Therapeutics* 274(3): 1399-1403.

Khan, Z. and H. A. Jinnah (2002). "Paroxysmal Dyskinesias in the Lethargic Mouse Mutant." *J. Neurosci.* 22(18): 8193-8200.

Kohl, M. M., O. Paulsen, et al. (2010). The Roles of GABAB Receptors in Cortical Network Activity. *Advances in Pharmacology*, Academic Press. Volume 58: 205-229.

Letts, V. A., R. Felix, et al. (1998). "The mouse stargazer gene encodes a neuronal Ca2+-channel [gamma] subunit." *Nat Genet* 19(4): 340-347.

Lin, F.-h., S. Barun, et al. (1999). "Decreased 45Ca2+ uptake in P/Q-type calcium channels in homozygous lethargic (Cacnb4lh) mice is associated with increased beta3 and decreased beta4 calcium channel subunit mRNA expression." *Molecular Brain Research* 71(1): 1-10.

Lin, F.-H., S. Lin, et al. (1999). "Glutamate decarboxylase isoforms in thalamic nuclei in lethargic mouse model of absence seizures." *Molecular Brain Research* 71(1): 127-130.

Lin, F.-h., Y. Wang, et al. (1995). "GABAB Receptor-Mediated Effects in Synaptosomes of Lethargic (lh/lh) Mice." *Journal of Neurochemistry* 65(5): 2087-2095.

Liu, H., M. De Waard, et al. (1996). "Identification of Three Subunits of the High Affinity omega-Conotoxin MVIIC-sensitive Ca2+ Channel." *Journal of Biological Chemistry* 271(23): 13804-13810.

Ludwig, A., V. Flockerzi, et al. (1997). "Regional Expression and Cellular Localization of the alpha1 and beta Subunit of High Voltage-Activated Calcium Channels in Rat Brain." *The Journal of Neuroscience* 17(4): 1339-1349.

McEnery, M. W., T. D. Copeland, et al. (1998). "Altered Expression and Assembly of N-type Calcium Channel alpha1B and beta Subunits in Epileptic lethargic(lh/lh) Mouse." *Journal of Biological Chemistry* 273(34): 21435-21438.

Mori, Y., M. Wakamori, et al. (2000). "Reduced Voltage Sensitivity of Activation of P/Q-Type Ca2+ Channels is Associated with the Ataxic Mouse MutationRolling Nagoya (tg rol)." *The Journal of Neuroscience* 20(15): 5654-5662.

Qian, J. and J. L. Noebels (2000). "Presynaptic Ca2+ Influx at a Mouse Central Synapse with Ca2+ Channel Subunit Mutations." *The Journal of Neuroscience* 20(1): 163-170.

Shirley, T. L., L. M. Rao, et al. (2008). "Paroxysmal dyskinesias in mice." *Movement Disorders* 23(2): 259-264.

Sidman, R. L., M. C. Green, et al. (1965). *Catalog of the Neurological Mutants of the Mouse.*

Slaght, S. n. J., N. Leresche, et al. (2002). "Activity of Thalamic Reticular Neurons during Spontaneous Genetically Determined Spike and Wave Discharges." *The Journal of Neuroscience* 22(6): 2323-2334.

Song, I., D. Kim, et al. (2004). "Role of the alpha1G T-Type Calcium Channel in Spontaneous Absence Seizures in Mutant Mice." *J. Neurosci.* 24(22): 5249-5257.

Steriade, M. and D. Contreras (1995). "Relations between cortical and thalamic cellular events during transition from sleep patterns to paroxysmal activity." *The Journal of Neuroscience* 15(1): 623-642.

Subramanyam, P., G. Obermair, et al. (2009). "Activity and calcium regulate nuclear targeting of the calcium channel beta4b subunit in nerve and muscle cells." *Channels* 3(5).

Talley, E. M., G. Solórzano, et al. (2000). "Low-voltage-activated calcium channel subunit expression in a genetic model of absence epilepsy in the rat." *Molecular Brain Research* 75(1): 159-165.

Tsakiridou, E., L. Bertollini, et al. (1995). "Selective increase in T-type calcium conductance of reticular thalamic neurons in a rat model of absence epilepsy." *The Journal of Neuroscience* 15(4): 3110-3117.

Volsen, S. G., N. C. Day, et al. (1997). "The expression of voltage-dependent calcium channel beta subunits in human cerebellum." *Neuroscience* 80(1): 161-174.

Weiergräber, M., U. Stephani, et al. (2010). "Voltage-gated calcium channels in the etiopathogenesis and treatment of absence epilepsy." *Brain Research Reviews* 62(2): 245-271.

Xu, X., Y. J. Lee, et al. (2011). "The Ca2+ Channel beta4c Subunit Interacts with Heterochromatin Protein 1 via a PXVXL Binding Motif." *Journal of Biological Chemistry* 286(11): 9677-9687.

Zhang, Y., M. Mori, et al. (2002). "Mutations in High-Voltage-Activated Calcium Channel Genes Stimulate Low-Voltage-Activated Currents in Mouse Thalamic Relay Neurons." *J. Neurosci.* 22(15): 6362-6371.

In: Calcium Channels: Properties, Functions and Regulation   ISBN: 978-1-61470-232-0
Editor: Mark R. Figgins   © 2012 Nova Science Publishers, Inc.

Chapter 7

ACTIVATION MECHANISMS OF STORE-OPERATED CALCIUM CHANNELS IN HUMAN NEUTROPHILS

Costantino Salerno[**] and Carlo Crifò

Department of Gynaecological Sciences, Perinatology and Children Health and
Department of Biochemical Sciences, University of Roma La Sapienza, Roma, Italy

Abstract

Calcium mobilization plays an important role in the regulation of superoxide anion secretion by neutrophils. Intracellular calcium increase is predominantly a result of calcium influx from extracellular media through the opening of calcium-permeable channels, which is subsequent to the emptying of intracellular stores. This capacitative mechanism, referred as store-operated calcium entry (SOCE), implies that depletion of agonist-sensitive intracellular calcium stores generates a secondary signal that promotes plasma membrane calcium influx. Accumulating evidence indicates that three protein families (TRPC, STIM, and Orai) play obligatory roles in the activation of this pathway by forming a ternary heterologous complex on the plasma membrane. This complex might mediate communication between the endoplasmic reticulum and the plasma membrane, perhaps by facilitating coupling between TRPC and inositol 3,4,5-trisposphate receptors. STIM1 behaves as a sensor of Ca^{2+} level in the endoplasmic reticulum, while Orai 1, by interacting with TRPC and STIM1, might act as a regulatory subunit that operates the transduction of the signals to the calcium-permeable channels on the plasma membrane. The role of the microtubule network in the regulation of SOCE is still obscure. Experiments performed with microtubule inhibitors gave rise to contradictory results, since these compounds induced partial depletion of Ca^{2+} stores and influx of bivalent cations from extracellular medium in the cytosol, but at the same time they inhibited SOCE triggered by agonists that are known to deplete Ca^{2+} from the endoplasmic reticulum.

[**] Correspondence author: Laboratory of Clinical Biochemistry, via dei Sardi 58, 00185 Roma, Italy.
Tel/Fax: +39 064463776; E-mail: costantino.salerno@uniroma1.it

1. INTRODUCTION

Neutrophils play a major role in host defence because they rapidly migrate to sites of infection and destroy invading microorganisms. The destruction of microorganisms may include nonoxidative and oxidative mechanisms. Nonoxidative killing involves the secretion of lysosomal enzymes. Oxidative-dependent killing is mediated by oxygen metabolites generated upon activation of the neutrophil NADPH oxidase. This enzyme is a multicomponent complex, including cytosolic and membrane-bound proteins, which is unassembled in resting cells. The cytosolic component includes four soluble factors (p67^{phox}, p47^{phox}, p40^{phox}, and a small G-protein Rac), while the membrane component consists of a stable, heterodimeric flavocytochrome (Cyt b$_{558}$). The agents used for neutrophil activation (opsonized particles, soluble inflammatory mediators) trigger different signal pathways, resulting in the organization and recruitment of cytosolic factors to plasma or phagosomal membranes, where NADPH oxidase is assembled [1-4].

Localization of superoxide anion production induced by opsonized *Escherichia coli* or zymosan particles differs from the one induced by soluble inflammatory mediators. Opsonized zymosan causes translocation of cytosolic components to the granule membrane. Activation of NADPH oxidase is triggered locally in the phagosome and, thus, oxygen metabolites are formed intracellularly. Global changes of intracellular Ca^{2+} resulting from Ca^{2+} influx are necessary but insufficient to activate NADPH oxidase during phagocytosis, since blockade of Ca^{2+} influx is accompanied by a slowed phagocytosis and a decrease of NADPH oxidase activity. Changes of intracellular Ca^{2+} level occur during β$_2$ integrin engagement at the point of contact between the particle and the cell and are subsequent to intracellular Ca^{2+} store release near the plasma membrane [5], but the identity of the intracellular messenger responsible for the release of these peripheral Ca^{2+} stores remains unknown. A global change of intracellular Ca^{2+} level occurs at the time of phagosomal closure and is temporally correlated with the activation of NADPH oxidase [6]. Phosphatidylinositol 3,4,5-trisphosphate (PIP$_3$) accumulation and anchoring at the phagocytic membrane are prerequisites for the generation of the Ca^{2+} signal [7]. It has been suggested [8] that PIP$_3$ may activate a phospholipase C isoform triggering inositol 3,4,5-trisposphate (InsP$_3$) production. It is also possible, as described in platelets [9], that PIP$_3$ regulates a novel pathway of Ca^{2+} entry, which is independent of an increase of phospholipase C activity and which would sustain NADPH oxidase activity during phagocytosis.

Interaction of soluble inflammatory mediators (such as f-Met-Leu-Phe, fMLP) with G-protein-coupled plasma membrane receptors results in the generation of InsP$_3$ by phospholipase Cβ, which causes the release of Ca^{2+} from intracellular stores by activating the channels located in the endoplasmic reticulum membranes [10]. Depletion of agonist-sensitive intracellular Ca^{2+} stores generates a secondary signal that promotes plasma membrane calcium influx. This mechanism, known as store-operated calcium entry (SOCE), results in the formation of a transient Ca^{2+} spike with a substantial amplitude, which returns to the basal level of Ca^{2+} concentration, dependent on the activity of the plasma membrane and endoplasmatic reticulum Ca^{2+}-ATPase pumps [11]. Activation of NADPH oxidase requires a second Ca^{2+}-independent signal, acting in synergy with Ca^{2+} influx from SOCE and leading to the phosphorylation of cytosolic phox proteins, which interact with the membrane component, and to the dissociation of Rho-GDI (guanine nucleotide dissociation inhibitor) from Rac that translocates independently to the membrane after exchange of GDP for GTP [2-4] (Figure 1).

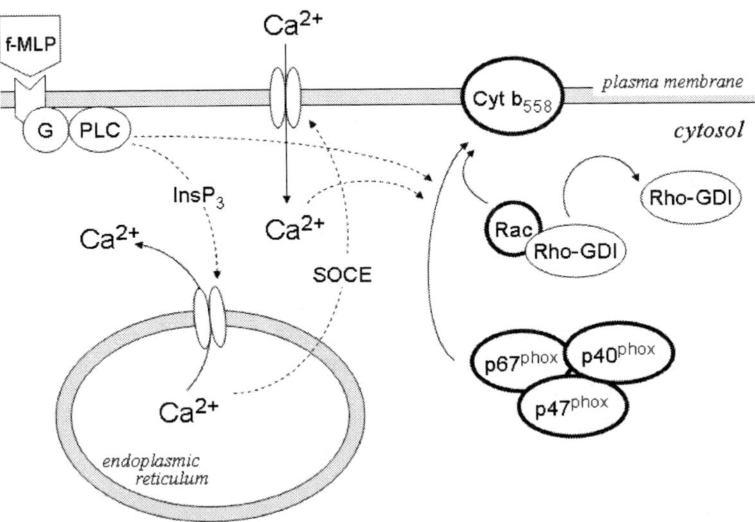

Figure 1. Proposed mechanism for activation of neutrophil NADPH oxidase by the soluble inflammatory peptide f-Met-Leu-Phe (f-MLP). In resting cells, NADPH oxidase components are compartmentalized between cell membranes and cytosol. The membrane component consists of a flavocytochrome (Cyt b$_{558}$), while the membrane component includes four soluble factors (p67^{phox}, p47^{phox}, p40^{phox}, and Rac). Interaction of f-MLP with G-protein-coupled plasma membrane receptors results in the generation of inositol 3,4,5-trisphosphate (InsP$_3$) by phospholipase Cβ (PLC), which activates the release of Ca^{2+} from intracellular stores. Calcium store depletion generates a secondary signal (SOCE) that promotes plasma membrane calcium influx. The transient increase in cytosolic Ca^{2+} level, acting in synergy with a second Ca^{2+}-independent signal, results in the binding of cytosolic phox proteins with the membrane NADPH-oxidase component and in the dissociation of Rho-GDI from Rac, which traslocates independently to the membrane.

Evidence for the requirement of extracellular Ca^{2+} entry for NADPH oxidase activation in plasma membrane is supported by a significant decrease of superoxide anion production when extracellular Ca^{2+} is suppressed or chelated by EGTA [12]. Many experiments, based on the use of specific pharmacological inhibitors, suggest that SOCE plays a role in the oxidative response triggered by soluble inflammatory mediators. The apparent link between SOCE and NADPH oxidase is supported by the relationship between modulation of Ca^{2+} entry and attenuation of NADPH oxidase activity in neutrophils that have been subjected to MRS1845, a selective SOCE inhibitor that, unlike other blockers, does not activate intracellular Ca^{2+} release at concentrations required to inhibit SOCE [13]. NADPH oxidase activity is also inhibited by the pyrazole derivative BTP2, another SOCE inhibitor, to the same level as in the absence of extracellular Ca^{2+} [14]. By contrast, elevation of periphagosomal Ca^{2+} is not affected by BTP2 and this confirms that SOCE does not partecipate in phagocytic stimuli-mediated Ca^{2+} influx.

2. STORE-OPERATED CALCIUM CHANNELS

The activation of Ca^{2+} entry through store-operated channels by agonists that deplete Ca^{2+} from the endoplasmic reticulum is a ubiquitous signal mechanism that allows the replenishment of intracellular calcium stores following a release event, thus maintaining the ability of the endoplasmic reticulum to release Ca^{2+} into the cytoplasm in response to subsequent Ca^{2+} releasing stimuli. Calcium ions that enter the cell via the store-operated pathway first access the cytoplasm before entering the endoplasmic reticulum and, in many cases, this results in sustained increase of cytoplasmic Ca^{2+} level, which is significant for some intracellular key processes [15]. Because of the role of SOCE in the regulation of many aspects of cell biology, several efforts have been made to understand the molecular structure of store-operated calcium channels. Accumulating evidence indicates that three protein families (TRPC, STIM, and Orai) play obligatory roles in the activation of this pathway (Figure 2).

It has been suggested that members of TRPC (transient receptor potential channel) family may be components of store-operated calcium channels [16]. Mammalian TRPC family consists of seven members that can be subdivided into subfamilies on the basis of their sequence homologies and functional similarities (TRPC1, TRPC2, TRPC4-5 and TRPC3-6-7 [17, 18]). In all members of the PRPC family, the cytoplasmic N-terminus and C-terminus are separated by six predicted transmembrane domains (TM1 to TM6) including a putative pore region between TM5 and TM6. The N-terminus is composed of three to four ankyrin repeats and a putative caveolin binding region, while the C-terminus includes a highly conserved proline rich motif. Human neutrophils express mRNAs for TRPC1, TRPC3, TRPC4, and TRPC6, and the corresponding proteins are found in the cell membrane [19]. It has been found [20, 21] that SOCE is enhanced by ectopic expression of TRPC1 or TRPC4 in several cell lines, while it is efficiently reduced by antisense constructs directed against TRPC1 or TRPC4, as well as by genetic disruption of the corresponding genes. It has been proposed [22] that TRPC1 is implicated in intracellular Ca^{2+} signalling by contributing to functional coupling between the plasma membrane and the endoplasmatic reticulum. Support for this hypothesis is derived from information obtained in human platelets and B lymphocytes, which suggests that TRPC1 is activated upon interaction with $InsP_3$ receptors on the endoplasmatic reticulum and that it acts not only as a component of the store-operated calcium channels, but also as a regulatory subunit of the $InsP_3$ receptors [23, 24].

Two STIM (stromal interaction molecule) homologues have been identified in mammalian tissues. These proteins are clearly closely related each other with respect to their primary amino acid sequence and their predicted secondary structure and domain organization. Both proteins are predicted to be single-pass transmembrane proteins with an exoplasmic N-terminal region and a cytoplasmic C-terminal region. STIM1 and STIM2 have almost identical EF-hand-containing N-terminal region and transmembrane domain, while their sequences deviate toward the C-terminus [25]. Suppressed STIM1 expression prevents SOCE, while over-expression of STIM1 induces only a small increase of Ca^{2+} influx, arguing that STIM1 may facilitate the assembly of store-operated calcium channels, without constituting the channel itself [26]. It has been suggested that STIM1 may function as a Ca^{2+} sensor or a coupler linking store depletion to activation of SOCE and that this function is mediated via the EF-hand Ca^{2+}-binding domain on the N-terminal portion of STIM1 inside the endoplasmic reticulum [27, 28]. Indeed, mutations in acidic residues within the Ca^{2+}-

binding pocket of the EF-band domain, which presumably lower the affinity for Ca^{2+}, produce constitutive Ca^{2+} entry, which is independent of store depletion, and a profound intracellular redistribution of STIM1 from a uniform pattern in the endoplasmic reticulum to spatially discrete areas termed puncta. Similar redistribution of STIM1 can be obtained by depleting endoplasmic-reticulum Ca^{2+} stores. Despite sharing close structural homology with STIM1, STIM2 has a very different role in the control of Ca^{2+} entry, since expression of STIM2 has a powerful inhibitory effect on SOCE activation in many cell lines (29).

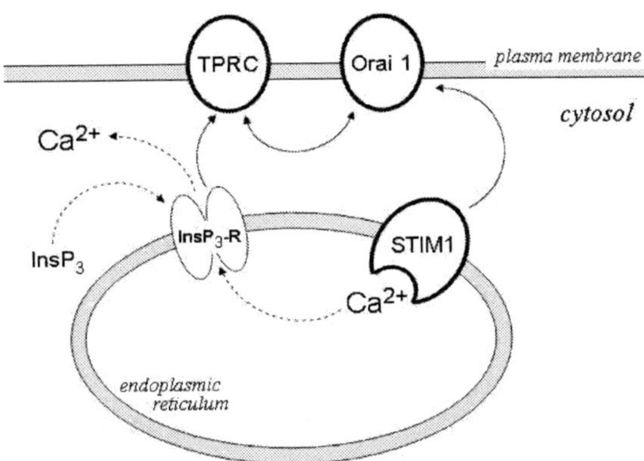

Figure 2. Potential role of STIM 1, Orai 1, and TRPC in the mechanisms linking Ca^{2+} depletion in endoplasmic reticulum to store-operated calcium channels. Upon internal Ca^{2+} store depletion, STIM 1 colocalyzes with Orai1 and TRPC in the plasma membrane to form a ternary complex that activates the transmembrane calcium channels. This reorganization mediates communication between the plasma membrane and the depleted intracellular calcium stores, facilitating the coupling between TPRC and inositol 3,4,5-trisphosphate receptors (InsP$_3$-R) on the endoplasmic reticulum.

Mammalian Orai family includes three homologues (Orai 1, Orai 2, and Orai 3) that are widely expressed at the mRNA level and can be incorporated into the plasma membrane when ectopically expressed [30]. Orai 1 is predicted to have four membrane spanning regions (M1 to M4) with its M3-M4 loop in the extracellular space and its N-terminus and C-terminus in the cytosol. The protein forms homodimers and homomultimers in cells and in detergent solutions, and can heteromultimerize with Orai 2 and Orai 3. Orai 1 is most likely an essential pore-forming subunit of the SOCE. Indeed, patients with one form of hereditary severe combined immune deficiency (SCID) and homozygous for a single missense mutation in Orai 1 are defective in store-operated calcium channels; expression of wild-type Orai 1 in SCID T cells restores store-operated Ca^{2+} influx [31]. Moreover, it has been found that SOCE is sensitive to mutation of two conserved acidic residues in the Orai-1 transmembrane segments: E106D and E190Q substitutions in transmembrane helices 1 and 3, respectively, diminish Ca^{2+} influx, increase current carried by monovalent cations, and render the channel permeable to Cs^{+} [32]. Several reports [27, 33] demonstrate that co-expression of Orai 1 and STIM1 results in a massive and rapid increase of SOCE. This increase is not seen when STIM1 alone is transfected into cells, while, when Orai 1 alone is transfected, there is a clear inhibition of SOCE [32]. It has been suggested [34] that Ca^{2+} release from intracellular stores promotes the binding of the STIM1 C-terminus with positively charged amino acid clusters that are present in the N-terminal cytosolic domain of Orai 1, reconstituting the transmembrane calcium channel.

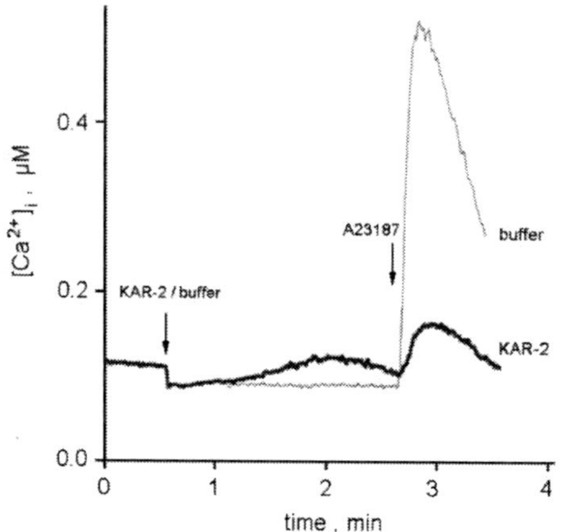

Figure 3. Time course of changes in intracellular free calcium concentration upon addition of semisynthetic bis-indol derivative (KAR-2) displaying anti-microtubular activity. Addition of 50 µM KAR-2 to calcium-refilled human neutrophil suspension in Ca^{2+}-free (EGTA-containing) medium caused a transient increase in cytosolic [Ca^{2+}], which was accompanied by a 70% reduction of calcium level in the intracellular Ca^{2+} stores. Cytosolic free calcium concentration was determined by measuring the emitted fluorescence of Fura-2 acetoxymethyl ester entrapped in the cells [44]. Calcium content into intracellular Ca^{2+} stores was determined by adding 1.6 µM 4-bromo-A23187 to the cell suspension as described by Montero et al. [45].

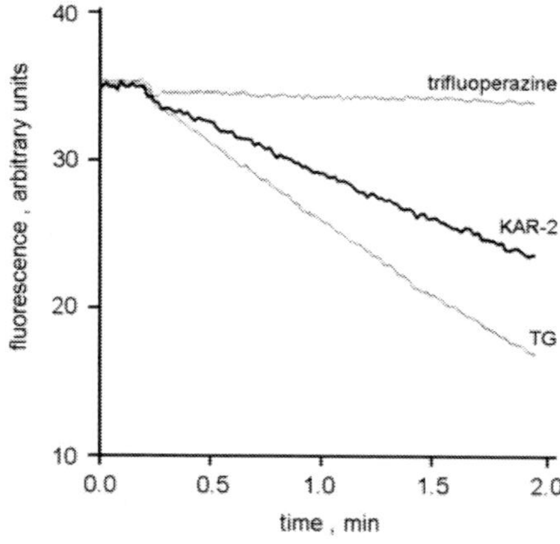

Figure 4. Time course of divalent cation uptake by calcium-depleted neutrophils. Magnesium ions, employed as Ca^{2+} surrogate, were taken up into the cells in the presence of 50 µM KAR-2. The uptake rate constant was about 40% of that observed in the presence of 8 nM thapsigargin (TG), that inhibits Ca^{2+} influx into the endoplasmic reticulum. As a control, the Mn^{2+} uptake was abolished by adding 25 µM trifluoperazine that inhibits calcium ion influx through the plasma membrane [46]. Other experimental conditions were as reported in Figure 3.

The precise location of STIM1 after store depletion is crucial in understanding its role in the reconstitution of the store-operated calcium channel. One proposal is that STIM1 is transported to and inserted into the plasma membrane upon Ca^{2+} store depletion [35]. Another possibility is that STIM1 aggregates underneath the plasma membrane without being inserted into the membrane [36]. A recent study reinforced the latter hypothesis by reporting little surface staining of EYFP (enhanced yellow fluorescent protein) antibody using an immunofluorescence method in HEK293 cells transfected with EYFP-STIM1 [37]. However, this result would be hard to reconcile with the electrophysiological experiments showing that extracellular application of antibody against the N-terminus of STIM1 blocks SOCE in HEK293 cells [26].

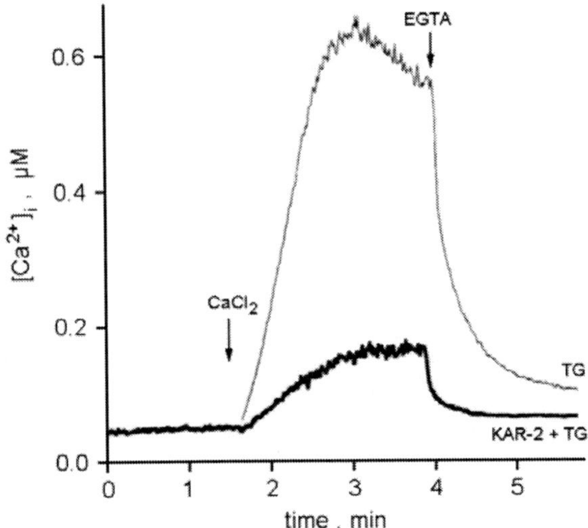

Figure 5. Inhibition of thapsigargin signal by KAR-2. When calcium-depleted neutrophils were treated with 8 nM thapsigargin (TG), addition of 0.5 mM $CaCl_2$ caused a dramatic increase in cytosolic calcium concentration because of the inhibition of calcium re-uptake by the endoplasmic reticulum, a condition that promotes calcium entry into the cells. This effect was markedly reduced by adding 50 µM KAR-2 that behaved as a SOCE inhibitor. Other experimental conditions were as reported in Figure 3.

3. ROLE OF CYTOSKELETON IN CALCIUM SIGNALING

Reversible actin polymerization is crucial for neutrophil development of polarity and migration, when the cells are exposed to a gradient of chemoattractants (chemotaxis) or to uniform concentrations of stimuli (chemokinesis) [38]. Position of the microtubule-organizing centre and of microtubules radiating from it is thought to enhance or maintain directional migration, in part by controlling vesicular traffic and membrane localization of cytoskeletal regulators [39]. In turn, actin-associated proteins at the cell cortex can enhance or stabilize depolymerization at the plus ends of microtubules, perhaps thereby designating destinations for delivery of regulatory molecules. The vectorial discharge of primary granules is dependent on calcium, but no evidence was found that calcium is involved in determining the polarity of exocytosis, while microtubules seem to contribute to the vectorial nature of the response [40].

Experiments performed in several cell lines (NIH 3T3, DDT$_1$MF-2, and A7r5 cells) unambiguously show that depolymerization of the actin cytoskeleton with cytochalasin does not negatively impact SOCE, thus suggesting that actin cytoskeleton does not play an obligate role per se in the store-operated calcium pathway [41, 42]. By contrast, disruption of microtubular cytoskeleton by nocodazole inhibits SOCE in HEK293 cells transfected with EYFP-STIM1 [37]. The inhibition is almost completely reversed by overexpression of EYFP-STIM1, implying that the inhibitory effect of SOCE by nocodazole is due to an impairment of STIM1 function. By using confocal microscopy analysis of the cells, it has been found that EYFP-STIM1 exhibits a fibrillar localization that colocalizes with endogenous α-tubulin. Depolarization of microtubules with nocodazole causes a change from a fibrillar EYFP-STIM1 localization to one that is similar to that of the endoplasmic reticulum. It is noteworthy that nocodazole alone is sufficient to activate SOCE in cells overexpressing EYFP-STIM1. It has been hypothesized that the organizing effect of the microtubule cytoskeleton on STIM1 localization and movements is required for optimal signalling when STIM1 is limiting or expressed at physiological levels, while it becomes unnecessary at very high levels of expression and even inhibits the access of STIM1 to the store-operated calcium channel.

The role of the neutrophil microtubule network in the regulation of SOCE has been studied by using a semisynthetic bis-indol derivative (KAR-2) with high anti-microtubular activity but virtually ineffective against calmodulin [43]. It has been found (Figs. 3, 4) that KAR-2 alone is sufficient to induce partial depletion of Ca^{2+} stores and influx of bivalent cations from extracellular medium in neutrophils purified from human blood, which presumably possess the physiological levels of neutrophil components of store-operated channels. Surprisingly, KAR-2 inhibits SOCE triggered by agonists that deplete Ca^{2+} from the endoplasmic reticulum (ATP, thapsigargin) while promoting Ca^{2+} influx in the cytoplasm on its own (Figure 5). As depolarization of microtubule cannot inhibit and activate SOCE in the same time, these findings suggest that, at least in neutrophils, microtubular cytoskeleton is engaged in the regulation of different mechanisms, which control Ca^{2+} traffic in the cell, and that some of them should be unrelated with the store-operated calcium pathway.

4. Conclusion

During the last decade, SOCE has been studied intensively and now we can rather definitively conclude that TRPC, STIM1 and Orai 1 are important actors in this process. Dissection of the mechanisms linking Ca^{2+} depletion in the endoplasmic reticulum to store-operated calcium channels might help to characterize the regulation of superoxide anion release and serve as targets for pharmacological drug development in inflammatory diseases. Many questions remain unanswered about the structural organization of the channels, the role of the microtubule network, and the mechanisms linking Ca^{2+} elevation to superoxide anion production. Although SOCE is considered the prominent mechanism for Ca^{2+} influx into neutrophils after Ca^{2+} pool discharge, it is becoming increasingly apparent that cell stimulation not only activates SOCE but also promotes additional Ca^{2+} entry pathways.

Identification of signalling molecules that govern distribution of intracellular Ca^{2+} will significantly enhance our understanding of neutrophil activation mechanisms.

REFERENCES

[1]   Groemping, Y. & Rittinger, K. (2005). Activation and assembly of the NADPH oxidase: a structural perspective. *Biochem. J.*, *386*, 401-416.

[2]   Sheppard, F. R., Kelher, M. R., Moore, E. E., McLaughlin, N. J., Banerjee, A. & Silliman, C. C. (2005). Structural organization of the neutrophil NADPH oxidase: phosphorylation and translocation during priming and activation. *J. Leukoc. Biol.*, *78*, 1025-1042.

[3]   Nobuhisa, I., Takeya, R., Ogura, K., Ueno, N., Kohda, D., Inagaki, F. & Sumimoto, H. (2006). Activation of the superoxide-producing phagocyte NADPH oxidase requires co-operation between the tandem SH3 domains of p47^{phox} in recognition of a polyproline type II helix and an adjacent alpha-helix of p22^{phox}. *Biochem. J.*, *396*, 183-192.

[4]   Ueyama, Y., Tatsuno, T., Kawasaki, T., Tsujibe, S., Shirai, Y., Sumimoto, H., Leto, T. L. & saito, N. (2007). A regulated adaptor function of p40^{phox}: distinct p67^{phox} membrane targeting by p40^{phox} and by p47^{phox}. *Mol. Biol. Cell*, *18*, 441-454.

[5]   Dewitt, S., Laffafian, I. & Hallett, M. B. (2003). Phagosomal oxidative activity during β_2 integrin (CR3)-mediated phagocytosis by neutrophils is triggered by a non-restricted Ca^{2+} signal: Ca^{2+} controls time not space. *J. Cell. Sci.*, *116*, 2857-2865.

[6]   Dewitt, S. & Hallett, M. B. (2002). Cytosolic Ca^{2+} changes and calpain activation are required for integrin-accelerated phagocytosis by human neutrophils. *J. Cell Biol.*, *159*, 181-189.

[7]   Dewitt, S., Tian, W. & Hallett, M. B. (2006). Localized PtdIns(3,4,5)P$_3$ or PtdIns(3,4)P$_2$ at the phagocytic cup is required for both phagosome closure and Ca^{2+} signalling in HL60 neutrophils. *J. Cell. Sci.*, *119*, 443-451.

[8]   Rameh, L. E., Rhee, S. G., Spokes, K., Kazlauskas, A., Cantley, L. C. & Cantley, L. G. (1998). Phosphoinositide 3-kinase regulates phospholipase Cγ-mediated calcium signalling. *J. Biol. Chem.*, *273*, 23750-23757.

[9]   Pasquet, J. M., Quek, L., Stevens, C., Bobe, R., Huber, M., Duronio, V., Krystal, G. & Watson, S. P. (2000). Phosphatidylinositol 3,4,5-trisphosphate regulates Ca^{2+} entry via btk in platelets and megakaryocytes without increasing phospholipase C activity. *EMBO J.*, *19*, 2793-2802.

[10]   Putney, J. W. (2007). Recent breakthroughs in the molecular mechanism of capacitative calcium entry (with thoughts on how we got here). *Cell Calcium*, *42*, 103-110.

[11]   Capuozzo, E., Verginelli, D., Crifò, C. & Salerno, C. (1997). Effects of calmodulin antagonists on calcium pump and cytosolic calcium level in human neutrophils. *Biochim. Biophys. Acta*, *1357*, 123-127.

[12] Foyouzi-Youssefi, R., Petersson, F., Lew, D. P., Krause, K. H. & Nusse, O. (1997). Chemoattractant-induced respiratory burst: increases in cytosolic Ca^{2+} concentrations are essential and synergize with a kinetically distinct second signal. *Biochem. J., 322,* 709-718.

[13] Lee, C., Xu, D. Z., Feketeova, E., Kannan, K. B., Fekete, Z., Deitch, E. A., Livingston, D. H. & Hauser, C. J. (2005). Store-operated calcium channel inhibition attenuates neutrophil function and postshock acute lung injury. *J. Trauma., 59,* 56-63.

[14] Steinckwich, N., Frippiat, J. P., Stasia, M. J., Erard, M., Boxio, R., Tankosic, C., Doignon, I. & Nüsse, O. (2007). Potent inhibition of store-operated Ca2+ influx and superoxide production in HL60 cells and polymorphonuclear neutrophils by the pyrazole derivative BTP2. *J. Leukoc. Biol., 81,* 1054-1064.

[15] Guo, R. & Huang, L. (2008). New insights into the activation mechanism of store-operated calcium channels: roles of STIM and Orai. *J. Zhejiang Univ. Sci., B 9,* 591-601.

[16] Worley, P. F., Zeng, W., Huang, G. N., Yuan, J. P., Kim, J. Y., Lee, M. G. & Muallem, S. (2007). TRPC channels as STIM1-regulated store-operated channels. *Cell Calcium, 42,* 205-211.

[17] Montell, C. (2005). The TRP superfamily of cation channels. *Sci. STKE,* re3.

[18] Lepage, P. K. & Boulay, G. (2007). Molecular determinants of TRP channel assembly. *Biochem. Soc. Trans., 35,* 81-83.

[19] Itagaki, K., Kannan, K. B., Singh, B. B. & Hauser, C. J. (2004). Cytoskeletal reorganization internalizes multiple transient receptor potential channels and blocks calcium entry into human neutrophils. *J. Immunol., 172,* 601-607.

[20] Salgado, A., Ordaz, B., Sampieri, A., Zepeda, A., Glazebrook, P., Kunze D. & Vaca, L. (2007). Regulation of the cellular localization and function of human transient receptor potential channel 1 by other members of the TRPC family. *Cell Calcium, 43,* 375-387

[21] Ambudkar, I. S., Ong, H. L., Liu, X., Bandyopadhyay, B. & Cheng, K. T. (2007). TRPC1: the link between functionally distinct store-operated calcium channels. *Cell Calcium, 42,* 213-223.

[22] Bréchard, S. & Tschirhart, E. J. (2008). Regulation of superoxide production in neutrophils: role of calcium influx. *J. Leukoc. Biol., 84,* 1223-1237.

[23] Rosado, J. A., Brownlow, S. L. & Sage, S. O. (2002). Endogenously expressed TRP1 is involved in store-mediated Ca^{2+} entry by conformational coupling in human platelets. *J. Biol. Chem., 277,* 42157-42163.

[24] Mori, Y., Wakamori, M., Miyakawa, T., Hermosura, M., Hara, Y., Nishida, M., Hirose, K., Mizushima, A., Kurosaki, M., Mori, E., Gotoh, K., Okada, T., Fleig, A., Penner, R., Iino, M. & Kurosaki, T. (2002). Transient receptor potential 1 regulates capacitative Ca^{2+} entry and Ca^{2+} release from endoplasmic reticulum in B lymphocytes. *J. Exp. Med., 195,* 673-681.

[25] Soboloff, J., Spassova, M. A., Tang, X. D., Hewavitharana, T., Xu, W. & Gill, D. L. (2006). Orai1 and STIM reconstitute store-operated calcium channel function. *J. Biol. Chem., 281,* 20661-20665.

[26]   Spassova, M. A., Soboloff, J., He, L. P., Xu, W., Dziadek, M. A. & Gill, D. L. (2006). STIM1 has a plasma membrane role in the activation of store-operated Ca^{2+} channels. *Proc. Natl. Acad. Sci.*, USA, *103*, 4040-4054.

[27]   Hewavitharana, T., Deng, X., Soboloff, J. & Gill, D. L. (2007). Role of STIM and Orai proteins in the store-operated calcium signalling pathway. *Cell Calcium, 43*, 173-183.

[28]   Li, Z., Lu, J., Xu, P., Xie, X., Chen, L. & Xu, T. (2007). Mapping the interacting domains of STIM1 and Orai 1 in Ca^{2+}-store-depletion-triggered Ca^{2+} channel activation. *J. Biol. Chem., 282*, 29448-29456.

[29]   Soboloff, J., Spassova, M. A., Hewavitharana, T., He, L. P., Xu, Johnstone, L. S., Dziadek, M. A. W. & Gill, D. L. (2006). STIM2 in an inhibitor of STIM1-mediated store-operated Ca^{2+} entry. *Curr. Biol., 16*, 1465-1470.

[30]   Gwack, Y., Srikanth, S., Feske, S., Cruz-Guilloty, F., Oh-hora, M., Neems, D. S., Hogan, P. G. & Rao, A. (2007). Biochemical and functional characterization of Orai proteins. *J. Biol. Chem., 282*, 16232-16243.

[31]   Feske, S., Gwack, Y., Prakriya, M., Srikanth, S., Puppel, S. H., Tanasa, B., Hogan, P. G., Lewis, R. S., Daly, M. & Rao, A. (2006). A mutation in Orai 1 causes immune deficiency by abrogating CRAC channel function. *Nature, 441*, 179-185.

[32]   Prakriya, M., Feske, S., Gwack, Y., Srikanth, S., Rao, A. & Hogan, P. G. (2006). Orai1 is an essential pore subunit of the CRAC channel. *Nature, 443*, 230-233.

[33]   Takahashi, Y., Murakami, M., Watanabe, H., Hasegawa, H., Ohba, T., Munehisa, Y., Nobori, K., Ono, K., Iijima, T. & Ito, H. (2007). Essential role of the N-terminus of murine Orai 1 in sore-operated Ca^{2+} entry. *Biochem. Biophys. Res. Commun., 356*, 45-52.

[34]   Huang, G. N., Zeng, W., Kim, J. Y., Yuan, J. P., Han, L., Muallem, S. & Worley, P. F. (2006). STIM1 carboxyl-terminus activates native SOC, I(crac) and TRPC1 channels. *Nat. Cell. Biol., 8*, 1003-1010.

[35]   Zhang, S. L., Yu. Y., Roos, J., Kozak, J. A., Deerinck, T. J., Ellisman, M. H., Stauderman, K. A. & Cahalan, M. D. (2005). STIM1 is a Ca^{2+} sensor that activates CRAC channels and migrates from Ca^{2+} store to the plasma membrane. *Nature, 437*, 902-905.

[36]   Wu, M. M., Buchanan, J., Luik, R. M. & Lewis, R. S. (2006). Ca^{2+} store depletion causes STIM1 to accumulate in ER regions closely associated with the plasma membrane. *J. Cell. Biol., 174*, 803-813.

[37]   Smyth, J. T., DeHaven, W. I., Bird, G. S. & Putney, J. W. (2007). Role of the microtubule cytoskeleton in the function of store-operated Ca^{2+} channel activator STIM1. *J. Cell. Sci., 120*, 3762-3771.

[38]   Chung, C. Y., Funamoto, S. & Firtel, R. A. (2001). Signaling pathways controlling cell polarity and chemotaxis. *Trends Biochem. Sci., 26*, 557-566.

[39]   Xu, J., Wang, F., VanKeymeulen, A., Rentel, M. & Bourne, H. R. (2005). Neutrophil microtubules suppress polarity and enhance directional migration. *Proc. Natl. Acad. Sci.*, USA, *102*, 6884-6889.

[40] Trapper, H., Furuya, W. & Grinstein, S. (2002). Localized exocytosis of primary (lysosomal) granules during phagocytosis: role of Ca^{2+}-dependent tyrosine phosphorylation and microtubules. *J. Immunol.*, *168*, 5287-5296.

[41] Ribeiro, C. M., Reece, J. & Putney, J. W. (1997). Role of the cytoskeleton in calcium signaling in NIH 3T3 cells. An intact cytoskeleton is required for agonist-induced $[Ca^{2+}]_i$ signaling, but not for capacitative calcium entry. *J. Biol. Chem.*, *272*, 26555-26561.

[42] Patterson, R. L., VanRossum, D. B. & Gill, D. L. (1999). Store-operated Ca^{2+} entry: evidence for a secretion-like coupling model. *Cell*, *98*, 487-499.

[43] Orosz, F., Vértessy, B. G., Salerno, C., Crifò, C., Capuozzo, E. & Ovádi, J. (1997). The interaction of a new anti-tumour drug, KAR-2 with calmodulin. *Br. J. Pharmacol*, *121*, 955-962.

[44] Grynkiewicz, G., Poenie, M. & Tsien, R. Y. (1985). A new generation of Ca^{2+} indicators with greatly improved fluorescence properties. *J. Biol. Chem.*, *260*, 3440-3450.

[45] Montero, M., Alvarez, J. & Garcìa-Sancho, J. (1992). Control of plasma-membrane Ca^{2+} entry by the intracellular Ca^{2+} stores. Kinetic evidence for a short-lived mediator. *Biochem. J.*, *288*, 519-525.

[46] Capuozzo, E., Verginelli, D., Crifò, C. & Salerno, C. (1997). Effects of calmodulin antagonists on calcium pump and cytosolic calcium level in human neutrophils. *Biochim. Biophys. Acta*, *1357*, 123-127.

In: Calcium Channels: Properties, Functions and Regulation   ISBN: 978-1-61470-232-0
Editor: Mark R. Figgins   © 2012 Nova Science Publishers, Inc.

Chapter 8

MEMBRANE TRANSPORT OF TOXIC METALS BY IONIC MIMICRY IN MAMMALIAN CELLS: WHERE DO CALCIUM CHANNELS FIT IN?

Carla Marchetti [tt]

Istituto di Biofisica, Consiglio Nazionale delle Ricerche,
16149 Genova, Italy, via De Marini, 6

Abstract

Cellular membranes are basically impermeable to ions and have developed specific pathways (transporters, channels or pumps) to facilitate metal translocation. These physiological carriers are not ideally selective and their specificity spectrum may include xenobiotic species, such as toxic metals whose availability in the environment has increased enormously with the onset of the industrial era. Competition between divalent endogenous and noxious metal ions at specific sites on membrane transport proteins and enzymes is referred to as "ionic mimicry".

In this chapter I will present some studies on the permeation mechanisms through mammalian cell plasma membranes of lead (Pb) and cadmium (Cd), two metals whose toxicity has been linked to their putative ability to mimic calcium (Ca) and zinc (Zn) at specific binding sites. Both metals can permeate through mammalian cell membranes taking advantage of different Ca channels, but, while Cd appears to take advantage mainly of the same pathways as Ca, Pb is also rapidly taken up in different cell types by passive transport systems that are distinct from Ca channels and independent of specific stimuli.

To further elucidate the role of voltage-dependent Ca channels (VDCC) in Cd uptake, we compared the effect of this metal in two Chinese hamster ovary (CHO) cell lines, a wild type and modified cell line, which was permanently transfected with an L-type VDCC. Both cultures were subjected to brief (30-60 min) exposure to 50-100 µM Cd in serum-free culture medium. Cell death was evident after 18-24 h with comparable features in both cell lines. Although VDCC represent a pathway of Cd entry and

---

[tt] Phone: 39010-6475578; Fax: 39010-6475500; Email marchetti@ge.ibf.cnr.it

participate in Cd-induced toxicity, as demonstrated by the effect of DHP modifiers, expression of L-type Ca channels is not sufficient to modify Cd accumulation and sensitivity to a toxicologically significant extent. This study confirmed that both Cd and Pb can take advantage of VDCC to permeate the membrane, but these transport proteins are not the only, and frequently not the most important, pathways of permeation.

ABBREVIATIONS

VDCC	voltage-dependent Ca channels
CHO	Chinese hamster ovary
CHOCα	CHOCα9β3α2/δ4
DHP	1,4-dihydropyridine
agatoxin	ω-agatoxin fraction IVA from *Agenelopsis aperta* venom

INTRODUCTION

Some metals, namely sodium (Na), potassium (K), calcium (Ca) and magnesium (Mg) are among the essential nutrients of living cells, but several metals can be fatal to mammalian cells even in tiny amounts. A common classification tends to sort these potentially toxic metals into two classes: those required by living organisms as essential micronutrients, such as cobalt, (Co), copper (Cu), iron (Fe), manganese (Mn), molybdenum (Mo) and zinc (Zn) and those devoid of any biological function and thus potentially toxic even at very low concentrations, such as cadmium (Cd), chromium (Cr), mercury (Hg) and lead (Pb). These elements are often referred to as "heavy metals", because of their large atomic weight and/or density, and their noxious effects as "heavy metal toxicity". However, this terminology is somewhat abused and recently Duffus (2002) has pointed out that it lacks a precise meaning and can be misleading. These substances will be referred to simply as "toxic metals" in this chapter.

The difference between toxic and non-toxic metal is hard to define. Even micronutrients can be detrimental to living organisms, when present in excessive levels and a refined equilibrium between deficient and toxic concentrations has to be maintained. This is particularly important for very specialized tissues, such as the brain, where metals induce oxidative damage and some of the essential micronutrients, such as Fe, Zn and Cu, have been implicated in etiology and development of different neurological and neurodegenerative diseases (Castellani et al., 2007; Smith et al., 2007). Less obviously, living organisms may find use for non-essential toxic metals in extreme conditions. An elegant example of unexpected biological function of Cd has been recently reported in marine diatoms (Xu et al., 2008).

Despite their potential catalytical activity, toxic metals represent a threat to mammalian cells, principally because their occurrence in air, soil and food has increased significantly as a consequence of recent human industrial activity and specialized cells and tissues have not developed the intrinsic ability to distinguish between physiologically essential and noxious ions. Recent research has gathered an increasing body of evidence for the interaction of toxic metals with intracellular components that lead to cellular injury and defense, but the mechanisms of transport of these metals and metal-containing species across plasma membranes remain to be fully characterized. A special case is transport of these toxicants across tight membrane barriers, such as gastrointestinal and blood-brain barrier, that have developed as protective mechanisms for sensitive organs (Bressler et al., 2007).

Most evidence indicates that toxic metals are capable of replacing or *mimicking* essential metals in the first or early step of transport and metabolism, but then are incapable of mediating subsequent vital functions (Clarkson, 1993; Bridges and Zalups, 2005). The toxic effects are then a consequence of both impairment of transport systems and accumulation into the cell.

Cd and Pb are two highly toxic metals whose harmfulness has been linked to their putative ability to mimic calcium (Ca) and zinc (Zn) at specific binding sites. Due to a close structural similarity to Ca, Cd binds Ca-binding proteins, such as calmodulin (Sutoo, 1994; Ouyang and Vogel, 1998), while Pb binds several intracellular proteins, such as protein kinase C (Markovac and Goldstein, 1988; Long et al., 1994; Tomsig and Suszkiw, 1995), Ca- and Zn- binding (Goldstein, 1993; Kern et al., 2000) and synaptic proteins (Bouton et al., 2001). Both Cd and Pb are potent blockers of voltage-dependent Ca channels (VDCC) by competing with Ca at a cation specific site on the protein; therefore it is not surprising that they could be driven into the cell by depolarization. Voltage-dependent and receptor-operated Ca channels are regarded as the major route of Cd and Pb entry into mammalian cells, although it has become increasingly evident how these channels are frequently not the main pathway of influx of these metals and work in coordination with other transporters.

VDCC mediate Cd influx in excitable cells (Hinkle et al., 1987; Hinkle et al., 1992; Shibuya and Douglas, 1992; Hinkle and Osborne, 1994), including mammalian neurons (Usai et al., 1999) and have been proposed to participate in Cd uptake also in cells from non-excitable tissues (Souza et al., 1997). In previous works of my laboratory, we showed that in certain cells Cd permeation occurs mainly through VDCC of the L-type (Usai et al., 1997; Gavazzo et al., 2005). However, in other studies, the presence of VDCC *per se* did not seem to enhance sensitivity to Cd; for example VDCC expressing (PC12) and non-expressing (PC18) cells showed comparable LD_{50} for Cd (Hinkle and Osborne, 1994) and the role of this pathway in the induction of cell death appears questionable.

The situation is even more complicated for Pb, whose ability to mimic Ca is more debatable. While Clarkson, quoting Simons (Simons, 1985), argues that "the chemical basis for Pb mimicking Ca is not obvious", because the ionic radii of Ca and Pb are "not particularly close" (Clarkson, 1993), Bridges and Zalups (2005) make the opposite assumption stating that "since the ionic radius of Pb is similar to that of Ca, it is possible that Pb mimic Ca at the site of Ca transporter". This contrasting opinion indicates how difficult it is to reach factual conclusions based only on chemical parameters. At odds with observations in other cell types (Simons and Pocock, 1987; Tomsig and Suszkiw, 1991; Legare et al., 1998), in cerebellar granule neurons the greatest part of depolarization-driven Pb influx takes place through DHP-insensitive VDCC, which are characteristic of neuronal tissue, and Pb is also rapidly taken up even in the absence of stimuli that open Ca-permeable channels (Mazzolini et al., 2001). All these observations downsize the importance of mimicking Ca in the transport of lead, although this toxic metal has been shown to interfere heavily with Ca- and Zn-dependent processes.

In this chapter, I will present some general considerations and experimental evidence concerning influx pathways of Pb and Cd across mammalian cell membranes and the relative contribution of VDCC in the uptake of these two toxic metals.

MATERIALS AND METHODS

Cell Culture

Granule cells were prepared from cerebella of 8-day-old Wistar rats as previously described (Usai et al., 1999), plated on 20mm poly-L-lysine coated glass coverslips and maintained in basal Eagle's culture medium, supplemented with 10% fetal calf serum, 100 µg/ml gentamicin and 25mM KCl, in a humidified 95% air/CO_2 atmosphere at 37°C. Cultures were treated with 10µM cytosine arabinoside from day 1 to minimize proliferation of non-neuronal cells. Experiments were carried out in cultures between 5 and 13 days in vitro.

Wild type Chinese hamster ovary (CHO) cells were obtained from American Type Culture Collection. CHO stably transfected with cDNA encoding for different subunits of voltage-dependent calcium channels (VDCC) were a generous gift of Franz Hofmann and Norbert Klugbauer (Institut für Pharmakologie und Toxikologie, Münich Germany). The cells used in this study are termed CHOCα9β3α2/δ4 (abbreviated CHOCα) to indicate that they express the α_{1C-b} subunit of VDCC, as well as β3 and α2/δ4 subunits from smooth muscle (Welling et al., 1993).

Both wild type CHO and CHOCα cells were maintained in DMEM medium (Sigma Chemical Co, St. Louis, MO, USA) supplemented with 10% bovine serum, in a 5% CO_2 humidified atmosphere at 37°C. For CHOCα cells, the culture medium was routinely supplemented with antibiotic geneticin G418 20µgr/ml. For the experiments, cells were plated in 12-multiwell trays with or without glass coverslips at a density of $2x10^5$ cells/ml.

Microscopy

Imaging of intracellular Cd and Pb was performed on monolayer cultures plated on glass coverslips. For confocal microscopy, cells were incubated with Oregon Green 488 Bapta-1, a fluorescent indicator, whose fluorescence intensity is increased upon binding Ca, Cd or Pb. Neurons were incubated in 6 µM of the cell-permeant AM ester form of the dye for 45 min at 37°C, and then washed several times with standard saline at room temperature. Other indicators, Fura2 and Indo1, were used as previously described (Usai et al., 1999; Mazzolini et al., 2001; Esposito et al., 2005). Cd and Pb were applied in the absence of Ca and terminated by the addition of the membrane-permeant metal chelator (N,N,N′,N′- tetrakis- 2-pyridyl methyl ethylenediamin, TPEN, 100 µm).

Cells were imaged using a confocal laser scanning microscope Nikon PCM2000 (Nikon Instr., Florence, Italy) with an oil immersion 100×objective (Na=1.3), coupled to a 50 µm confocal pinhole condition (Diaspro et al., 1999).

Cd Treatment

Cells were subjected to 30-60 min treatments with 50 or 100 µM CdCl$_2$ in culture medium without serum. When specified, the medium was supplemented with 30 mM KCl to depolarize the cell membrane and with the dihydropyridines (DHPs) BayK 8644 or nimodipine, which are a L-type VDCC agonist and antagonist, respectively. Stock solutions of nimodipine and BayK were made up at 10 mM in 100% ethanol and diluted in medium to the final concentration. High KCl alone, as well as DHPs (no Cd added), had no effect on cell viability and appearance. After incubation cells were washed three times with PBS containing 1 mM EDTA and fresh medium containing serum was replaced. Cultures were kept in the incubator for another 18-24 hours before testing. Cell viability in each sample was assessed by trypan-blue exclusion assay. All chemicals, culture media and sera were from Sigma-Aldrich Italy.

Capacitance and Current Measurements by Patch-Clamp Technique

Membrane currents were measured from wild type CHO and permanently transfected CHOCα cells in whole-cell clamp configuration by a patch-clamp Axopatch amplifier (Molecular Devices Corporation, Union City, CA). Electrodes were manufactured from borosilicate glass capillaries (Hilgenberg GmbH, Malsfeld, Germany) and had resistance of ≈ 4 MΩ. Voltage stimulation and data acquisition were performed by a PC through a Digidata 1440A interface and Pclamp-10 software (Molecular Devices). Currents were low-pass filtered at 5 kHz and digitized at 10 kHz. Capacitance transients were minimized by analog compensation and the value obtained by this compensation was taken as an estimate of the cell capacitance. All currents traces were further corrected for leak and residual transients by a computer generated P/4 protocol. The holding potential was -80 mV in all the experiments. Current traces were analyzed with Clampfit-10 and Sigma Plot (Jandel Scientific, Erkrath, Germany) software.

Cells were continuously superfused by gravity flow (10 ml/min) with a solution containing (in mM) NaCl 140, KCl 5.4, CaCl$_2$ 1.8, Hepes 5. The pH was 7.4. The internal (pipette) solution contained (in mM): CsCl 20, CsOH 110, Aspartic acid 100, EGTA 5, Hepes 5, with pH adjusted to 7.3 with Trizma base. Calcium currents were recorded in similar external solution containing 130 mM NaCl and 5 mM CaCl$_2$. Wild type CHO cells do not possess prominent voltage-dependent currents, and CHOCα only contained the L-type VDCC under study; therefore there was no need to antagonize or minimize the other conductances to resolve Ca currents. Substitution of internal cesium (Cs) with equimolar potassium (K) had no appreciable effect on the measurements, confirming that CHO cells do not possess any voltage-dependent conductance, including that due to K permeable channels. Application of modifiers, such as agonist and antagonist DHPs, was accomplished by gravity flow; control ion substitution experiments showed that the external bath was completely changed in 10 sec, which was the maximal stimulation rate in these experiments

Apoptosis Detection

The occurrence of apoptosis was evaluated in living cells by Annexin V-FITC apoptosis detection kit (Biovision Research Products, Mountain View ,CA, USA). Monolayer cultures on glass coverslips were incubated with a solution of Annexin V-FITC (0.5 µg/ml), which binds to phosphatidylserine, and propidium iodide (1 µg/ml), which permeates into dead cells, in a buffer containing 10 mM Hepes/NaOH (ph 7.4), 140 mM NaCl and 2.5 mM $CaCl_2$. For evaluation of apoptotic nuclei, cells were incubated with 4,6-diamidino-2-phenylindole (DAPI, Sigma) 1µg/mL in PBS for 5-10 min, as in previous work (Gavazzo et al., 2005). Cells were observed on a Zeiss Axiovert 100 equipped with filters for UV and visible excitation. Nuclei were judged apoptotic when showing segmentation and masses mostly associated with the nuclear envelope or apoptotic bodies. Photographs were collected by a digital camera (Nikon Coolpix 995).

Absorption Spectroscopy Cd Determination

Cells were exposed to Cd in culture medium without added serum. After treatment, cultures were washed three times with PBS containing 1 mM EDTA and harvested immediately. Viable cells were counted in triplicate in each sample. Cells were then washed again three times with the above buffer, re-suspended in distilled water and disrupted by sonication (Sonopuls Ultrasonic Homogenizer, Bandelin) for 2 minutes in an ice bath. The total Cd content of each sample was measured by Flameless Atomic Absorption Spectroscopy (FAAS) using a Perkin Elmer Spectrophotometer (Model 1100 B) equipped with a graphite furnace (Model HGA 700), as in previous work (Gavazzo et al., 2005). The Cd content was normalized to the volume occupied by the cells and molarity was calculated (see results).

Statistical Analysis

Experiments were run in triplicate. Data are presented as mean ± standard error of the mean in at least two experiments. Statistical significance was evaluated by Student-Newman-Keuls multiple comparison test (In Stat, GraphPad Software).

RESULTS

Influx of Toxic Metals into Mammalian Cells

Influx of toxic divalent metal ions into mammalian cells can be monitored in cells pre-loaded with specific fluorescent probes, such as Fura2 (Hinkle et al., 1992; Usai et al., 1997; Usai et al., 1999; Mazzolini et al., 2001), Indo1 (Kerper and Hinkle, 1997a; Esposito et al., 2005), or Oregon green. These fluorescent indicators were developed and are predominantly used for determining intracellular calcium levels, but they have relatively high affinity also for other polyvalent ions. These include Cd and Pb, whose pathways of permeation have been studied by this technique.

Figure 1 shows how exposure to a solution containing a toxicologically significant concentration of Cd or Pb causes a sizeable increase in Oregon Green fluorescence, similar to that seen following Ca influx (Pellistri et al., 2004). This demonstrates feasibility of this probe as toxic metal influx indicator. In this experiment, Cd uptake was triggered by depolarization, indicating the essential contribution of VDCC, while neurons rapidly accumulate Pb even in the absence of a specific stimulus.

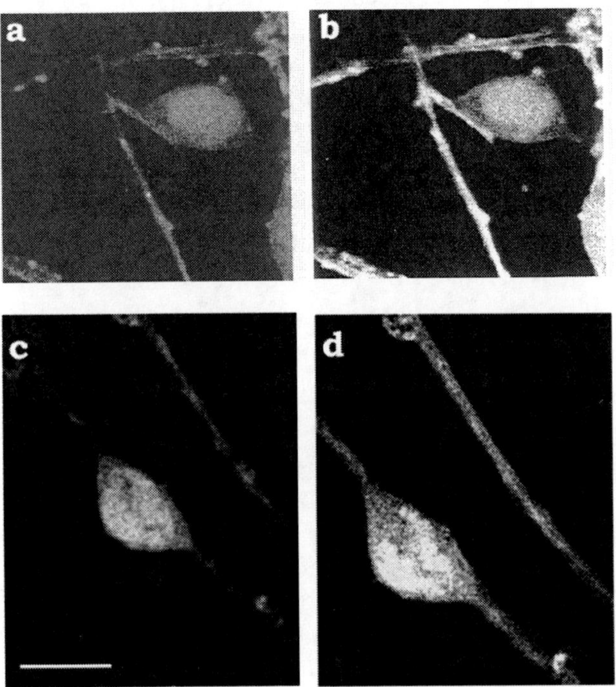

Figure 1. Confocal microscopy images of Cd (a, b) and Pb (c, d) uptake by cerebellar granule cells pre-loaded with the divalent metal-sensitive dye Oregon Green. In (a) and (c), neurons were bathed in a physiological saline. In (b) they have been superfused with a solution containing 0 Ca, 30 mM KCl and 100 µM Cd Cl_2, which caused the dye fluorescence to increase significantly. In (d), external solution contained 0 Ca and Pb 10 µM, which permeates through the neuron membrane even in the absence of a depolarizing stimulus and also caused an increase in the dye fluorescence. Bar = 10 µM.

This observation is in agreement with previous findings. In Fura2 loaded neurons (Usai et al., 1999) and insulinoma cells (Usai et al., 1997; Gavazzo et al., 2005), influx of Cd and Pb was evaluated from the fluorescence emission ratio R (E_{380}/E_{340}) and its rate of rise dR/dt. In Ca-free, high-KCl solution, R increased upon application of 100 µM Cd and experiments demonstrated that Cd permeates into these two cell types using Ca-selective pathways, and, in particular, through L-type VDCC. In contrast, in cerebellar neurons, Pb^{2+} (5 - 50µM) determined an increase of R even in the absence of any specific stimulus that open VDCC (Mazzolini et al., 2001; Esposito et al., 2005). In addition, although in Pb dR/dt was increased up to a factor of 5 by depolarizing high-KCl solution, suggesting permeation through VDCC, the effect was neither antagonized by nimodipine, nor enhanced by BayK8644; instead dR/dt decreased in the presence of the Q-type VDCC blocker ω-agatoxin IVA (Mazzolini et al., 2001), suggesting an involvement of non-L type VDCC. Further experiments also proved that glutamate channels of the NMDA type are an additional pathway of both Cd and Pb uptake (Usai et al., 1999; Mazzolini et al., 2001).

Pb caused a rapid saturation of the dye, whose intracellular concentration is ~ 10 μM, indicating that intracellular Pb^{2+} can readily reach concentration in the micromolar range. Parallel experiments showed that non-excitable HeLa cells take up Pb^{2+} much more slowly (Mazzolini et al., 2001), so fast uptake of Pb appears to be a peculiar feature of neurons and may be linked to specific lipidic component of neuronal membranes.

The study of Pb pathways of permeation in rat cerebellar granule neurons was further pursued by taking advantage of a different fluorescent dye, Indo1 (Kerper and Hinkle, 1997a; Esposito et al., 2005). Pb binds Indo1 with high affinity acting as a quencher. Its permeation through the neuronal membrane was indicated by a decrease of the fluorescence emission, which occurred even in resting condition confirming stimulus-independent permeation. Indo1 proved more suitable than Fura2 to study Pb permeation in neurons, because it accumulates into the cell to a concentration estimated 10 times larger than that of Fura2 and, due to a lower affinity for Pb, it is saturated by higher concentration of Pb. In 20 μM Pb, uptake reached a plateau level (≈ 45% of initial fluorescence) in 4 min and was partially antagonized by 25 μM lanthanum (La). Subsequent addition of a membrane permeant ionophore caused a further (> 70%) quenching of the dye, suggesting that previous saturation was due to inactivation of the transport system. Intracellular Pb concentrations were evaluated by comparing the fluorescence intensity of pre-loaded cells with that of cells individually loaded by patch pipette. This estimate indicated that the concentration of free Pb^{2+} sufficient to inactivate the transport system is close to 50 pM (Esposito et al., 2005), a dose of Pb that has been shown to induce a significant effect on the activity of intracellular proteins, such protein kinase C (Markovac and Goldstein, 1988; Long et al., 1994; Tomsig and Suszkiw, 1995), Ca-binding (Goldstein, 1993; Kern et al., 2000) and synaptic proteins(Bouton et al., 2001).

Role of VDCC in Cd-Induced Mortality

Given the prominent role of L-type VDCC in Cd uptake by excitable cells, the involvement of these channels in Cd toxicity was investigated further. Tests of mortality, chromatine condensation and DNA fragmentation showed that a pulsed Cd treatment induces delayed apoptotic cell death in VDCC-containing insulinoma cells and that nimodipine protects against Cd-induced apoptosis and necrosis in these cells, supporting a specific involvement of L-type VDCC (Gavazzo et al., 2005). The same treatments were largely harmless in VDCC-free HeLa cell cultures, in which neither death nor DNA fragmentation was observed. Therefore, excitable cells appear more susceptible than non-excitable epithelial-like cells to Cd accumulation. However, it is not clear whether the mere presence of VDCC can make cells more vulnerable to toxic metal injury. To answer this question we have compared the effects of Cd in two cell types that are virtually identical, except for the expression of VDCC in one of the two. This approach is made possible by the availability of a CHO cell line that has been permanently transfected with VDCC α_1, β and α_2/δ subunits, $CHOC\alpha9\beta_3\alpha_2/\delta_4$ cells (Welling et al., 1993; Cataldi et al., 1999).

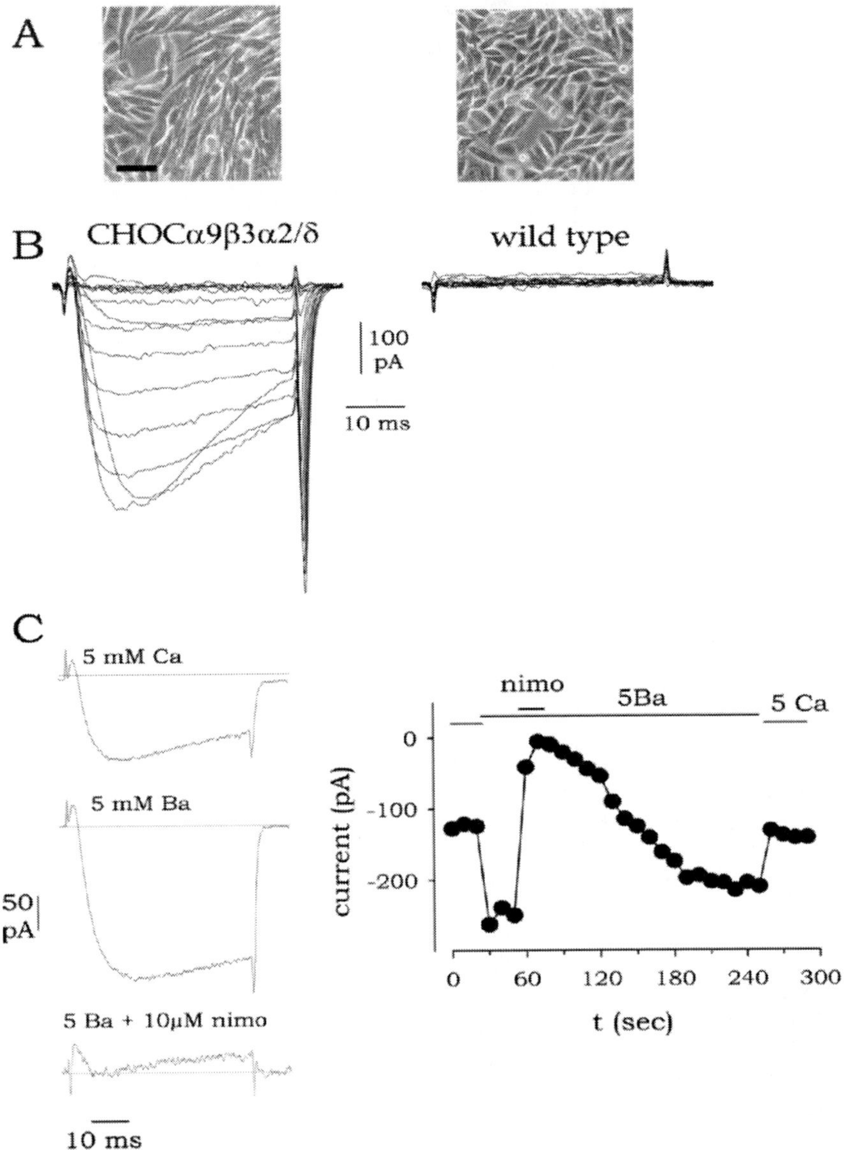

Figure 2. CHOCα cells express L-type VDCC .

(A) Representative microphotographs of monolayer cell cultures of permanently transfected CHOCα cells (left) and wild type CHO cells (right). Bar = 35 μM.

(B) Representative current traces evoked by voltage steps of 40 ms duration from –60 to +60 mV, in 10 mV intervals, from a holding potential of -80 mV in a CHOCα cell (left) and a wild type CHO cell (right). The external solution contained 5 mM $CaCl_2$. This protocol evoked voltage-dependent calcium current in the permanently transfected cell, while no current was present in CHO cell of the wild type.

(C) Characterization of the Ca current in CHOCα cell. The current was increased by more than 50% when the external solution was changed from 5 mM $CaCl_2$ to 5 mM $BaCl_2$ and it was reversibly blocked by 10 μM nimodipine. Current traces evoked by 50 ms depolarizing steps from –80 mV (holding potential) to +10 mV in the three conditions are shown on the left. The graph on the right shows the time course of the experiment.

Prominent voltage-dependent Ca currents were recorded from the permanently transfected cell line, while such currents are completely absent in wild type CHO (figure 2A). The currents are larger with Ba as charge carrier and are sensitive to DHP modulation (figure 2B), as expected for a L-type Ca channel.

Both cell types (wild type and CHOCα) were subjected to a 'pulse treatment' with 50 or 100 μM Cd in serum-free culture medium and observed after 18-24 hour after wash. Cell suffering and mortality was negligible immediately after the treatment, but it was evident after at least 16 hours. As previously observed (Galan et al., 2001; Poliandri et al., 2003; Gavazzo et al., 2005), Cd accumulates into the cell during treatment, but the toxic mechanism leading to cell death becomes effective at a later time. Figure 3 shows the appearance of the cell monolayer 24 hours after an incubation of 30 min with 100 μM Cd. Cells exposed to Cd were of smaller size than control cells and had an increased tendency to detach from their neighbors and from the substrate and float. In addition, CHOCα VDCC-expressing cells frequently acquired an irregular shape, with shredded edges (figure 3). The number of viable cells was estimated from the number of adherent, trypan-blue excluding cells and this number was normalized to that of adherent viable cells in control conditions, as in our previous works (Gavazzo et al., 2005). Despite some morphological features, the dose dependence of Cd sensitivity was very similar in the two cell types (figure 3). Therefore, by this approach it is not possible to demonstrate a clear role of VDCC in mediating Cd-induced cell detachment.

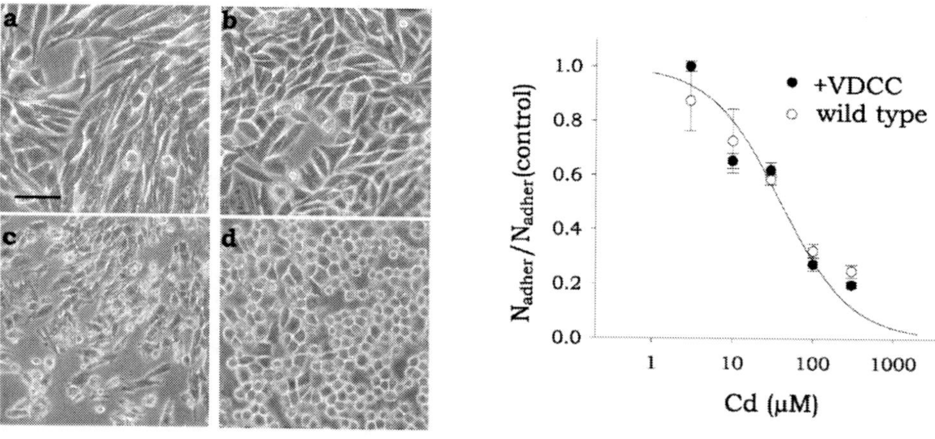

Figure 3. Effect of pulse treatment with Cd in CHOCα VDCC-expressing and wild type CHO cells. Microphotographs of control (a, b) and Cd-treated (c, d) cells: (a, c) VDCC-expressing CHOCα cells and (b, d) wild type CHO cells. Both cell types were incubated in 100 μM Cd for 60 minutes in the presence of 30 mM KCl. Pictures were taken 24 h after wash of the metal. Bar = 35 μM.

The graph shows the effect of a 60 min pulse treatment with Cd on cell adhesion, as a function of concentration. Trypan-blue excluding adherent cells were counted 24 h after wash of the metal. Points are average ± sem of 3 experiments in the same condition in CHOCα (filled circles) and wild type CHO (empty circles) and were best fitted to the function

$$N_{adher}/N_{adher(control)} = 1/(1+([Cd]/ ED_{50}))$$

where $N_{adher}/N_{adher(control)}$ is the number of adherent cells after Cd treatment normalized to the number of adherent cells in control culture; [Cd] is the concentration of Cd and ED_{50} is the concentration of Cd that causes detachment from the substrate of 50% of cells. The best fit yielded ED_{50} =40 µM for CHOCα and 43 µM for wild type CHO. The two curves are overlapped. In contrast with the different appearance, the two cell types were similarly affected by Cd treatment.

In subsequent experiments, we challenged the cells with Cd in the presence of 1 µM DHP modifiers. When VDCC-expressing CHOCα cells were exposed to Cd in the presence of the antagonist (nimodipina) or agonist (BayK 8644), the effect of the metal was clearly modified with respect to cells exposed to Cd alone (figure 4), with a sizeable increase of mortality in the presence of BayK and protection in the presence of nimodipine. DHP modifiers were largely ineffective in wild-type CHO cells exposed to Cd (not shown). Therefore, it seems that the actual contribution of L-type VDCC to Cd uptake is clearly underscored only in the presence of these modifiers.

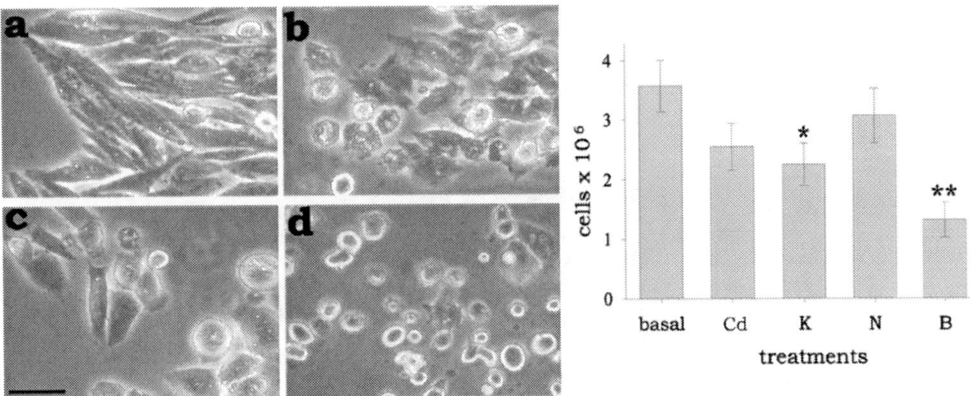

Figure 4. Effect of dihydropyridines on Cd cytotoxicity in VDCC-expressing CHOCα cells. VDCC-expressing CHOCα cells were treated for 30 min in (a) control medium containing 30 mM KCl, (b) 100 µM Cd + 30 mM KCl ; (c) 100 µM Cd + 30 mM KCl + 1µM nimodipine; (d) 100 µM Cd + 30 mM KCl + 1µM BayK8644. Note the partial recovery in shape of the cells treated in the presence of nimodipine and definitive lost of adhesion when the treatment was performed in the presence of BayK. Bar = 25 µM.

The graph represents counts of viable adherent cells in the same experiment (average ± sem in 3 samples) in control (basal), following for a 30 min treatment with 100 µM Cd (Cd), Cd+30 mM KCl (K), Cd+30 mM KCl and 1 µM nimodipine (N), Cd+30 mM KCl and 1 µM BayK (B). * indicates significantly different from control with p < 0.05, and ** with p < 0.0001

In the attempt to reveal any discrepancy in Cd sensitivity between wild type and VDCC-expressing CHO cells, we investigated other features of cellular death caused by transient acute exposure to Cd. Previous work has shown how Cd-induced mortality is basically due to apoptosis in several cell types (el Azzouzi et al., 1994; Hamada et al., 1997; Chrestensen et al., 2000; Lopez et al., 2003) and in CHO cells in particular (Banfalvi et al., 2005). In insulinoma cells, we have shown that apoptosis is the main pathway of cell demise following short (0.5 – 1 hour) treatments with relatively low doses of metal (≤ 100 µM; (Gavazzo et al., 2005). We investigated development of Cd-induced apoptosis by two indicators: exposure of phosphatidylserine, measured by Annexin V binding (Martin et al., 1995) and chromatine condensation (Banfalvi et al., 2005). Although both redistribution of phospholipids and chromatin condensation are regarded as early events in the apoptotic process, we were unable to identify any of these markers before 16 hours after treatment in all the cultures. Therefore initiation of apoptosis did not occur before that time, or before that time it is not possible to foresee the occurrence of apoptosis by straightforward means. In CHO cells that had been treated with 100 µM Cd for one hour and tested for apoptosis occurrence 18-24 hours later, incubation with annexin V revealed green staining in the plasma membrane, indicating translocation of phosphotidylserine to the outer surface of the membrane (figure 5 a-d). Less frequently, green cells also displayed a red staining throughout the nucleus (figure 5 e), indicating loss of membrane integrity, a feature of late apoptosis or necrosis. Figure 6 shows living cells stained with DAPI to visualize nuclear morphology. In cells treated with medium only, nuclei appeared homogeneously bright (figure 6a,b). In cells that had been exposed to Cd, nuclei were smaller, frequently condensed and segmented, as typical of apoptotic cells (figure 6c,d). Again, despite some variation in appearance, counts of stained cells could not reveal any statistically significant difference and reveal a specific sensitivity of CHOCα cells to Cd-induced apoptosis.

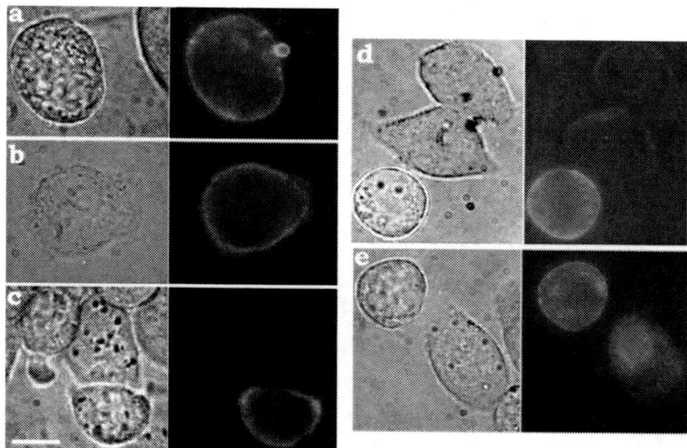

Figure 5. Evidence of Cd-induced apoptosis: membrane staining. Cells were treated with 100 µM Cd for 30 minutes and, 24h after wash of the metal, they were incubated with Annexin V-FITC (0.5 µg/ml), and propidium iodide (1 µg/ml) for 15-45 min. In each panel, the left side is the bright field photography of the cell and the right side is a fluorescent image obtained using a dual filter set for FITC and rhodamine. Panel a through d are from CHOCα cells, while e is from wild type CHO. In a, b and c, exposure of phosphatidylserine on the outer leaflet of the plasma membrane is revealed by the green staining; no red staining was detectable in these cells; in d the green staining is more diffuse into the cells and a pale red staining is also present; in e both cells have lost membrane integrity and taken up the red dye (PI) and are in late apoptotic (top cell) and necrotic state (bottom right cell).

Another indicator of apoptosis is cell shrinkage, which can be quantified by cell capacitance measurements. Both wild type and CHOCα cells showed a decrease in cell capacitance 18-24 hours after Cd treatment. In wild type, cell capacitance was 24 ± 2 pF (n= 15) in control and 15 ± 3 pF (n=6) after incubation with 100 µM Cd, significantly different from control with p< 0.05. In CHOCα cells, cell capacitance was 35 ± 3 pF (n= 17) in control and 22 ± 2 pF (n =39) after incubation with 100 µM Cd, significantly different from control with p< 0.001. In CHOCα cells, the effect of Cd treatment was partially prevented by nimodipine and the difference between this condition and control condition was not significant (figure 7). This indicates again that the contribution of L-type VDCC in Cd-induced toxicity is evident only in the presence of DHPs.

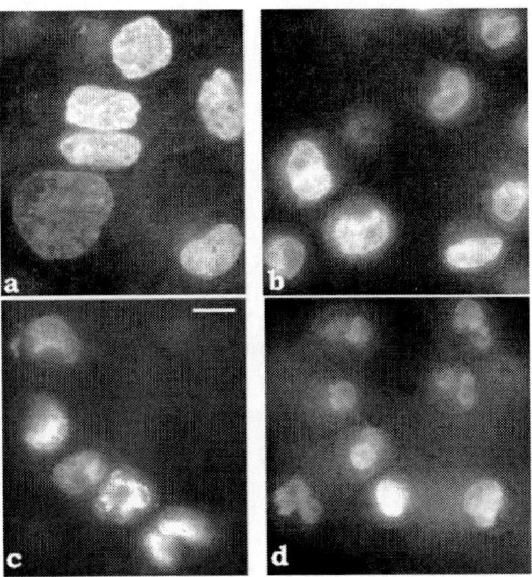

Figure 6. Evidence of Cd-induced apoptosis: nuclear staining. CHOCα (a, c) and wild type CHO (b, d) cells were incubated with medium only (a, b) or 100 µM Cd with 30 mM KCl (c, d) for 30 minutes and stained with DAPI 24h after wash of the metal. Bar = 12 µM. Both cell types were reduced in size by Cd treatment and their nuclei were condensed and segmented.

Finally, we performed measurements of total Cd accumulation during incubation time by atomic absorption spectroscopy. For these experiments, cells were harvested immediately after the treatment and the total amount of Cd was normalized to the cell volume, estimated from the average single cell capacitance. Assuming that both cell types are approximately spherical and have a specific membrane capacitance of $1 \mu F/cm^2$, 10^6 wild type CHO cells comprise a volume of 11µl, while 10^6 CHOCα cells a volume of 19.5 µl. This difference is not negligible and, with this method, results were considerably more reproducible than using a standard normalization to the total protein content. Data shown in figure 8 indicate that the difference in Cd accumulation is barely or not significant when both cell types are incubated in 100 µM Cd with or without elevated KCl. On the other hand, BayK 8644 significantly enhanced metal accumulation in depolarized CHOCα cells treated with 50 or 100 µM Cd, but not in wild type CHO cells, while nimodipine caused a significantly decrease in total metal content during incubation with 50 µM Cd in the permanently transfected cells.

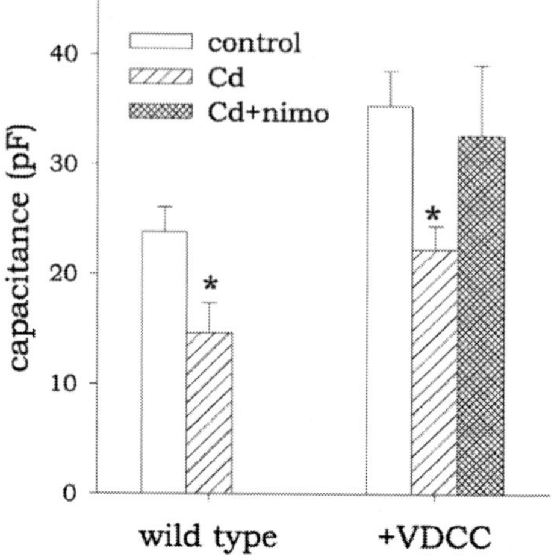

Figure 7. Evidence of Cd-induced apoptosis: change in cell capacitance. Cells were incubated with 100 µM Cd for 60 minutes and electrical measurements were performed 24h after wash. Membrane capacitance was estimated from transient compensation (see methods section). Treatment with Cd caused shrinkage of all cell, as revealed by reduction of the cell capacitance in both wild-type and CHOCα cells. * indicates a significant difference from control with $p < 0.05$. Nimodipine protected CHOCα cells from Cd-induced shrinkage.

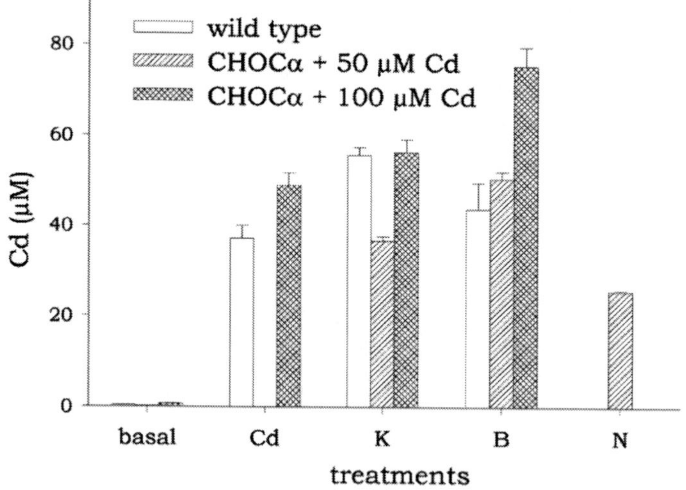

Figure 8.Total Cd accumulation measured by FAAS in wild type CHO and CHOCα cells. Both cell types were incubated for 1 hour in medium containing 0, 50 or 100 µM Cd and other modifiers. Cd determination was normalized to the cell volume (see text). Treatments were as follows: control 0 Cd (basal), 60 min Cd (Cd), Cd+30 mM KCl (K), Cd+30 mM KCl and 1 µM BayK (B), Cd+30 mM KCl and 1 µM nimodipine (N). Depolarization (treatment K) did not increase significantly the intracellular concentration of Cd, with respect to Cd alone ($p > 0.05$). In CHOCα cells, but not in wild type CHO, BayK significantly enhanced and nimodipine significantly reduced Cd accumulation, both with $p < 0.001$, with respect to the same dose of Cd administered alone in depolarization (treatment K).

DISCUSSION

Toxicity of both Cd and Pb is mainly due to their ability to permeate not simply mammalian cell membranes, but specifically tight membrane barriers, such as the epithelial lining in the gastrointestinal tract, luminal membrane of proximal tubules in the kidney and the blood-brain barrier. In early work, the influx of Pb and Cd into mammalian cells was frequently ascribed to permeation through calcium channels, and in particular L-type DHP-sensitive VDCC (Hinkle et al., 1987; Simons and Pocock, 1987; Tomsig and Suszkiw, 1991; Hinkle and Osborne, 1994; Souza et al., 1997). More recently other types of VDCC, different from the classical DHP-sensitive L-type, have been implicated both in Cd (Usai et al., 1999; Leslie et al., 2006) and Pb (Mazzolini et al., 2001) uptake.

In the attempt to identify the clear role of VDCC in toxic metal uptake, it is remarkable that Pb permeation appears to be largely independent of VDCC even in neurons (Mazzolini et al., 2001; Esposito et al., 2005). As far as Cd is concerned, in this chapter I have shown that expression of L-type VDCC is not sufficient to modify Cd uptake to a toxicological significant extent, at least in the case of a brief, intense exposure, as those used in this study. Cd accumulation and sensitivity were comparable in two cell lines whose only difference was the presence of L-type VDCC in one of the two. Indeed, wild-type CHO cells, which do not possess VDCC, are vulnerable to Cd poisoning and take up Cd during relatively brief exposures to an extent sufficient to trigger cell death by apoptosis or necrosis similarly to VDCC expressing CHOCα cells. Although VDCC play an important role in Cd uptake in CHOCα cells, as indicated by the sizeable effect of L-type channel DHP modifiers, CHO cells lacking VDCC must use other transport proteins with similar efficiency and outcome.

It is thus clear that VDCC are not the main responsible of toxic metal permeation through cell membranes and tight barriers. Rapid passive transport of Pb, independent of VDCC, was reported at the brain endothelium (Deane and Bradbury, 1990) and even protein-independent transport of lead has been described (Diaz and Monreal, 1995). Store-depletion activated calcium channels have been suggested to mediate the uptake of Pb in various cells, including a glial cell line (Kerper and Hinkle, 1997a; Legare et al., 1998) and brain capillary endothelial cells (Kerper and Hinkle, 1997b). In a rat astroglial cell line, Cheong et al. (2004) reported kinetics data concerning two putative distinct pH-sensitive transport mechanisms for Pb. VDCC-independent uptake of Pb by cerebellar granule neurons was inhibited by La (Esposito et al., 2005) and this observation can provide some clue as for the mechanism of permeation. The involvement of a La-sensitive store-depletion activated channel, as proposed in other systems (Kerper and Hinkle, 1997a; Evans et al., 2003), is unlikely because Pb influx did not require drainage of the stores. Another possibility is that Pb^{2+} enters the neuron via reverse operation of an exchanger similar (or identical) to the Na-Ca exchanger, which is also very sensitive to La block (Kaczorowski et al., 1989) and whose role in Zn permeation was demonstrated in cortical neurons (Sensi et al., 1997). Also this possibility was discarded because Na-Ca exchanger does not play a significant role when the Na concentrations are close to the physiological value, as in Esposito et al. (2005). It is more probable that Pb is taken up through a La-sensitive ion carrier similar to the Zn transporter found in brain neurons (Colvin, 1998), whose identity and features still wait full definition.

Involvement of other transport proteins has been described. At the epithelial lining of the gastrointestinal barrier, DMT1, the main transporter that absorbs Fe in the brush border membrane of the mammalian intestine has been shown to carry also different toxic metals (Picard et al., 2000; Lecoeur et al., 2002; Bannon et al., 2003; Okubo et al., 2003; Suzuki et al., 2007). The role of this transporter in Cd absorption is underscored by the fact that iron deficiency creates a significant risk for increased cadmium exposure by increasing gastrointestinal absorption from 5% to as much as 20%. In general, a low level of essential metals favor uptake of non-essential (toxic) elements that compete for their site and this stresses the importance of evaluating metal balance, in contrast with metal concentration. In addition, other metal-specific transporters have been identified at the cDNA level (Bressler et al., 2007) and studied in expression systems (Liu et al., 2008).

Research has also driven attention to the importance of active back transport, mediated by active transporter and pumps, which mitigates metal uptake. This role is frequently played by ATP-driven metal pumps (P_{1B}-type ATPases) that actively transport several metals, including toxicant, from a variety of cells (Solioz and Vulpe, 1996; Arguello et al., 2007). Interestingly, a functional link has been postulated between L-type VDCC and the Zn transporter ZnT-1, which would attenuate Cd and Zn influx thus limiting accumulation (Ohana et al., 2006). This is in agreement with other studies that showed how the expression of ZnT-1 inhibits the activity of L-type channels (Beharier et al., 2007) and is a notable example of how different transport systems may act in a concerted manner in the regulation of metal transport.

In the balance between positive and negative factors, the expression of detoxification proteins is also an important component, chiefly in the case of Cd, as revealed by the presence of a strong negative correlation between cellular MT content and the rate of apoptosis induced by Cd (Shimoda et al., 2001).

The present study completes and corrects our previous work (Gavazzo et al., 2005). Data also confirm that short treatments with Cd trigger delayed cell death, chiefly by apoptosis. The latency time is longer than 8 hours, and during this time it was not possible to find any indication of the forthcoming demise. It is probable that the cell, once hit by the toxic insult, tries all available defense strategies to face the insult, until it reaches a point of no return. The time course of this process would in principle be dependent on the noxious stimulus intensity and cell susceptibility. However, this latency could not be shortened even increasing Cd concentration up to 300 µM in any of the model cells investigated (Gavazzo et al., 2005), an observation that requires further study.

CONCLUSION

Ionic mimicry is a useful framework to study the mechanisms of metal toxicity. Uptake of toxic metals, such as Cd and Pb, by mammalian cells of different tissues occurs through many different pathways, of which VDCC have frequently been regarded as prominent. Because of the abundance of VDCC, excitable cells may be more susceptible to accumulate Cd and Pb than non-excitable cells. In this chapter I have shown that although Cd and Pb can permeate through VDCC, this is not the most relevant mechanism of passive uptake of Pb, while the mere presence of VDCC is not sufficient to make cells significantly more vulnerable to Cd injury. Other factors, including different transport systems and expression of detoxification proteins, may be more relevant in the accumulation process of these toxic metals.

ACKNOWLEDGMENTS

The experiments described in this chapter were carried out by the author with the collaboration of several people. Ilaria Zanardi and Irena Baranowska-Bosiacka helped in Cd treatment experiments, while Elisabetta Morelli performed the FAAS measurements. I thank Mauro Robello and Paola Gavazzo for help and suggestions throughout this work. I am also grateful to Franz Hofmann and Norbert Klugbauer (Institut für Pharmakologie und Toxikologie, Münich Germany) for their permission to use the permanently transfected CHO cells and to Maurizio Taglialatela and Mauro Cataldi (Università Federico II, Naples, Italy) for sending me the cells.

REFERENCES

Arguello, J. M., Eren, E. and Gonzalez-Guerrero, M. (2007). The structure and function of heavy metal transport P1B-ATPases. *Biometals, 20,* 233-48.

Banfalvi, G., Gacsi, M., Nagy, G., Kiss, Z. B. and Basnakian, A. G. (2005). Cadmium induced apoptotic changes in chromatin structure and subphases of nuclear growth during the cell cycle in CHO cells. *Apoptosis, 10,* 631-42.

Bannon, D. I., Abounader, R., Lees, P. S. and Bressler, J. P. (2003). Effect of DMT1 knockdown on iron, cadmium, and lead uptake in Caco-2 cells. *Am. J. Physiol. Cell Physiol, 284,* C44-50.

Beharier, O., Etzion, Y., Katz, A., Friedman, H., Tenbosh, N., Zacharish, S., Bereza, S., Goshen, U. and Moran, A. (2007). Crosstalk between L-type calcium channels and ZnT-1, a new player in rate-dependent cardiac electrical remodeling. *Cell Calcium, 42,* 71-82.

Bouton, C. M., Frelin, L. P., Forde, C. E., Arnold Godwin, H. and Pevsner, J. (2001). Synaptotagmin I is a molecular target for lead. *J. Neurochem, 76,* 1724-35.

Bressler, J. P., Olivi, L., Cheong, J. H., Kim, Y., Maerten, A. and Bannon, D. (2007). Metal transporters in intestine and brain: their involvement in metal-associated neurotoxicities. *Hum. Exp. Toxicol, 26,* 221-9.

Bridges, C. C. and Zalups, R. K. (2005). Molecular and ionic mimicry and the transport of toxic metals. *Toxicol Appl. Pharmacol, 204,* 274-308.

Castellani, R. J., Moreira, P. I., Liu, G., Dobson, J., Perry, G., Smith, M. A. and Zhu, X. (2007). Iron: the Redox-active center of oxidative stress in Alzheimer disease. *Neurochem. Res, 32,* 1640-5.

Cataldi, M., Secondo, A., D'Alessio, A., Taglialatela, M., Hofmann, F., Klugbauer, N., Di Renzo, G. and Annunziato, L. (1999). Studies on maitotoxin-induced intracellular Ca(2+) elevation in chinese hamster ovary cells stably transfected with cDNAs encoding for L-type Ca(2+) channel subunits. *J. Pharmacol. Exp. Ther., 290,* 725-30.

Cheong, J. H., Bannon, D., Olivi, L., Kim, Y. and Bressler, J. (2004). Different mechanisms mediate uptake of lead in a rat astroglial cell line. *Toxicol. Sci., 77,* 334-40.

Chrestensen, C. A., Starke, D. W. and Mieyal, J. J. (2000). Acute cadmium exposure inactivates thioltransferase (Glutaredoxin), inhibits intracellular reduction of protein-glutathionyl-mixed disulfides, and initiates apoptosis. *J. Biol. Chem.*, *275*, 26556-65.

Clarkson, T. W. (1993). Molecular and ionic mimicry of toxic metals. *Annu. Rev. Pharmacol. Toxicol.*, *33*, 545-71.

Colvin, R. A. (1998). Characterization of a plasma membrane zinc transporter in rat brain. *Neurosci. Lett*, *247*, 147-50.

Deane, R. and Bradbury, M. W. (1990). Transport of lead-203 at the blood-brain barrier during short cerebrovascular perfusion with saline in the rat. *J. Neurochem.*, *54*, 905-14.

Diaspro, A., Annunziata, S., Raimondo, M., Ramoino, P. and Robello, M. (1999). A single-pinhole confocal laser scanning microscope for 3-D imaging of biostructures. *IEEE Eng. Med. Biol. Mag.*, *18*, 106-10.

Diaz, R. S. and Monreal, J. (1995). Protein-independent lead permeation through myelin lipid liposomes. *Mol. Pharmacol.*, *47*, 766-71.

Duffus, J. H. (2002). "Heavy metals" a meaningless term? *Pure Appl. Chem.*, *74*, 793-807.

el Azzouzi, B., Tsangaris, G. T., Pellegrini, O., Manuel, Y., Benveniste, J. and Thomas, Y. (1994). Cadmium induces apoptosis in a human T cell line. *Toxicology*, *88*, 127-39.

Esposito, A., Robello, M., Pellistri, F. and Marchetti, C. (2005). Two-photon analysis of lead accumulation in rat cerebellar granule neurons. *Neurochem. Res.*, *30*, 949-54.

Evans, T. J., James-Kracke, M. R., Kleiboeker, S. B. and Casteel, S. W. (2003). Lead enters Rcho-1 trophoblastic cells by calcium transport mechanisms and complexes with cytosolic calcium-binding proteins. *Toxicol. Appl. Pharmacol*, *186*, 77-89.

Galan, A., Troyano, A., Vilaboa, N. E., Fernandez, C., de Blas, E. and Aller, P. (2001). Modulation of the stress response during apoptosis and necrosis induction in cadmium-treated U-937 human promonocytic cells. *Biochim. Biophys. Acta*, *1538*, 38-46.

Gavazzo, P., Morelli, E. and Marchetti, C. (2005). Susceptibility of insulinoma cells to cadmium and modulation by L-type calcium channels. *Biometals*, *18*, 131-42.

Goldstein, G. W. (1993). Evidence that lead acts as a calcium substitute in second messenger metabolism. *Neurotoxicology*, *14*, 97-101.

Hamada, T., Tanimoto, A. and Sasaguri, Y. (1997). Apoptosis induced by cadmium. *Apoptosis*, *2*, 359-67.

Hinkle, P. M., Kinsella, P. A. and Osterhoudy, K. C. (1987). Cadmium uptake and toxicity via voltage-sensitive calcium channels. *J.Biol.Chem.*, *262*, 16333-16337.

Hinkle, P. M. and Osborne, M. E. (1994). Cadmium toxicity in rat pheochromocytoma cells: studies on the mechanism of uptake. *Toxicol. Appl. Pharmacol*, *124*, 91-8.

Hinkle, P. M., Shanshala, E. D. and Nelson, E. J. (1992). Measurements of intracellular cadmium with fluorescent dyes - Further evidence for the role of calcium channels in cadmium uptake. *J.Biol.Chem.*, *267*, 25553-25559.

Kaczorowski, G. J., Slaughter, R. S., King, V. F. and Garcia, M. L. (1989). Inhibitors of sodium-calcium exchange: identification and development of probes of transport activity. *Biochim. Biophys. Acta*, *988*, 287-302.

Kern, M., Wisniewski, M., Cabell, L. and Audesirk, G. (2000). Inorganic lead and calcium interact positively in activation of calmodulin. *Neurotoxicology, 21*, 353-63.

Kerper, L. E. and Hinkle, P. M. (1997a). Cellular uptake of lead is activated by depletion of intracellular calcium stores. *J. Biol. Chem, 272*, 8346-52.

Kerper, L. E. and Hinkle, P. M. (1997b). Lead uptake in brain capillary endothelial cells: activation by calcium store depletion. *Toxicol. Appl. Pharmacol, 146*, 127-33.

Lecoeur, S., Huynh-Delerme, C., Blais, A., Duche, A., Tome, D. and Kolf-Clauw, M. (2002). Implication of distinct proteins in cadmium uptake and transport by intestinal cells HT-29. *Cell. Biol. Toxicol, 18*, 409-23.

Legare, M. E., Barhoumi, R., Hebert, E., Bratton, G. R., Burghardt, R. C. and Tiffany-Castiglioni, E. (1998). Analysis of Pb2+ entry into cultured astroglia. *Toxicol. Sci, 46*, 90-100.

Leslie, E. M., Liu, J., Klaassen, C. D. and Waalkes, M. P. (2006). Acquired cadmium resistance in metallothionein-I/II(-/-) knockout cells: role of the T-type calcium channel Cacnalpha1G in cadmium uptake. *Mol. Pharmacol, 69*, 629-39.

Liu, Z., Li, H., Soleimani, M., Girijashanker, K., Reed, J. M., He, L., Dalton, T. P. and Nebert, D. W. (2008). Cd2+ versus Zn2+ uptake by the ZIP8 HCO3--dependent symporter: kinetics, electrogenicity and trafficking. *Biochem. Biophys. Res. Commun, 365*, 814-20.

Long, G. J., Rosen, J. F. and Schanne, F. A. (1994). Lead activation of protein kinase C from rat brain. Determination of free calcium, lead, and zinc by 19F NMR. *J. Biol. Chem, 269*, 834-7.

Lopez, E., Figueroa, S., Oset-Gasque, M. J. and Gonzalez, M. P. (2003). Apoptosis and necrosis: two distinct events induced by cadmium in cortical neurons in culture. *Br. J. Pharmacol., 138*, 901-11.

Markovac, J. and Goldstein, G. W. (1988). Picomolar concentrations of lead stimulate brain protein kinase C. *Nature, 334*, 71-3.

Martin, S. J., Reutelingsperger, C. P., McGahon, A. J., Rader, J. A., van Schie, R. C., LaFace, D. M. and Green, D. R. (1995). Early redistribution of plasma membrane phosphatidylserine is a general feature of apoptosis regardless of the initiating stimulus: inhibition by overexpression of Bcl-2 and Abl. *J. Exp. Med, 182*, 1545-56.

Mazzolini, M., Traverso, S. and Marchetti, C. (2001). Multiple pathways of Pb(2+) permeation in rat cerebellar granule neurones. *J. Neurochem, 79*, 407-16.

Ohana, E., Sekler, I., Kaisman, T., Kahn, N., Cove, J., Silverman, W. F., Amsterdam, A. and Hershfinkel, M. (2006). Silencing of ZnT-1 expression enhances heavy metal influx and toxicity. *J. Mol. Med, 84*, 753-63.

Okubo, M., Yamada, K., Hosoyamada, M., Shibasaki, T. and Endou, H. (2003). Cadmium transport by human Nramp 2 expressed in Xenopus laevis oocytes. *Toxicol. Appl. Pharmacol., 187*, 162-7.

Ouyang, H. and Vogel, H. J. (1998). Metal ion binding to calmodulin: NMR and fluorescence studies. *Biometals, 11*, 213-22.

Pellistri, F., Cupello, A., Esposito, A., Marchetti, C. and Robello, M. (2004). Two-photon imaging of calcium accumulation in rat cerebellar graule cells. *Neuroreport, 15*, 83-87.

Picard, V., Govoni, G., Jabado, N. and Gros, P. (2000). Nramp 2 (DCT1/DMT1) expressed at the plasma membrane transports iron and other divalent cations into a calcein-accessible cytoplasmic pool. *J. Biol. Chem, 275*, 35738-45.

Poliandri, A. H., Cabilla, J. P., Velardez, M. O., Bodo, C. C. and Duvilanski, B. H. (2003). Cadmium induces apoptosis in anterior pituitary cells that can be reversed by treatment with antioxidants. *Toxicol Appl. Pharmacol, 190*, 17-24.

Sensi, S. L., Canzoniero, L. M., Yu, S. P., Ying, H. S., Koh, J. Y., Kerchner, G. A. and Choi, D. W. (1997). Measurement of intracellular free zinc in living cortical neurons: routes of entry. *J. Neurosci, 17*, 9554-64.

Shibuya, I. and Douglas, W. W. (1992). Calcium channels in rat melanotrophs are permeable to manganese, cobalt, cadmium, and lanthanum, but not to nickel: evidence provided by fluorescence changes in fura-2-loaded cells. *Endocrinology, 131*, 1936-41.

Shimoda, R., Nagamine, T., Takagi, H., Mori, M. and Waalkes, M. P. (2001). Induction of apoptosis in cells by cadmium: quantitative negative correlation between basal or induced metallothionein concentration and apoptotic rate. *Toxicol. Sci, 64*, 208-15.

Simons, T. J. (1985). Influence of lead ions on cation permeability in human red cell ghosts. *J. Membr. Biol., 84*, 61-71.

Simons, T. J. and Pocock, G. (1987). Lead enters bovine adrenal medullary cells through calcium channels. *J. Neurochem., 48*, 383-9.

Smith, D. G., Cappai, R. and Barnham, K. J. (2007). The redox chemistry of the Alzheimer's disease amyloid beta peptide. *Biochim. Biophys. Acta, 1768*, 1976-90.

Solioz, M. and Vulpe, C. (1996). CPx-type ATPases: a class of P-type ATPases that pump heavy metals. *Trends Biochem. Sci, 21*, 237-41.

Souza, V., Bucio, L. and Gutierrez-Ruiz, M. C. (1997). Cadmium uptake by a human hepatic cell line (WRL-68 cells). *Toxicology, 120*, 215-20.

Sutoo, D. (1994). Disturbance of brain function by exogenous cadmium. The Vulnerable Brain and Environmental Risks. R. L. Isaacson and K. F. Jensen. New York, Plenum Press. **3**: 281-299.

Suzuki, T., Momoi, K., Hosoyamada, M., Kimura, M. and Shibasaki, T. (2007). Normal cadmium uptake in microcytic anemia mk/mk mice suggests that DMT1 is not the only cadmium transporter in vivo. *Toxicol Appl. Pharmacol, 17*, 17.

Tomsig, J. L. and Suszkiw, J. B. (1991). Permeation of Pb2+ through calcium channels: fura-2 measurements of voltage- and dihydropyridine-sensitive Pb2+ entry in isolated bovine chromaffin cells. *Biochem.Biophys.Acta, 1069*, 197-200.

Tomsig, J. L. and Suszkiw, J. B. (1995). Multisite interactions between Pb2+ and protein kinase C and its role in norepinephrine release from bovine adrenal chromaffin cells. *J. Neurochem, 64*, 2667-73.

Usai, C., Barberis, A., Moccagatta, L. and Marchetti, C. (1997). Pathways of cadmium uptake in excitable mammalian cells: a microspectrofluorimetric study. *Eur. J. Histochem, 41*, 189-90.

Usai, C., Barberis, A., Moccagatta, L. and Marchetti, C. (1999). Pathways of cadmium influx in mammalian neurons. *J. Neurochem*, *72*, 2154-61.

Welling, A., Bosse, E., Cavalie, A., Bottlender, R., Ludwig, A., Nastainczyk, W., Flockerzi, V. and Hofmann, F. (1993). Stable co-expression of calcium channel alpha 1, beta and alpha 2/delta subunits in a somatic cell line. *J. Physiol*, *471*, 749-65.

Xu, Y., Feng, L., Jeffrey, P. D., Shi, Y. and Morel, F. M. (2008). Structure and metal exchange in the cadmium carbonic anhydrase of marine diatoms. *Nature*, *452*, 56-61.

INDEX

A

absorption, 175, 178
absorption spectroscopy, 175
acid, 154, 155, 167
activation, x, 151, 152, 153, 154, 159, 160, 161, 181
active transport, 178
acute, 160, 173
adhesion, 173
aggregates, 157
agonist, x, 151, 152, 162, 167, 173
air, 164, 166
alpha, 183
amplitude, 152
analog, 167
anemia, 182
anhydrase, 183
antagonist, 167, 173
anterior pituitary, 182
antibiotic, 166
antibody, 157
antioxidants, 182
antisense, 154
apoptosis, 168, 170, 173, 174, 175, 176, 177, 178, 180, 181, 182
apoptotic, 168, 170, 174, 179, 182
application, 169
arabinoside, 166
astroglial, 177, 179
atmosphere, 166
attention, 178
availability, x, 163, 170

B

barrier, 177
barriers, 165, 177
beta, 183

binding, x, 153, 154, 155, 163, 165, 166, 170, 174, 180, 181
biological, 164
blocks, 157, 160
blood, 158, 165, 177, 180
borosilicate glass, 167
bovine, 166, 182
brain, 164, 177, 179, 180, 181, 182
buffer, 168

C

cadmium, x, 163, 164, 178, 179, 180, 181, 182, 183
calcium, x, 151, 152, 153, 154, 155, 156, 157, 158, 159, 160, 161, 162, 163, 164, 165, 166, 168, 171, 177, 179, 180, 181, 182, 183
calmodulin, 165, 181
capacitance, 167, 174, 175, 176
capillary, 177, 181
carrier, 172, 177
cation, 156, 160, 165, 182
cations, 182
cerebellar granule cells, 169
cerebrovascular, 180
cesium, 167
channels, x, 151, 152, 154, 155, 158, 160, 161, 163, 164, 165, 167, 170, 178, 182
chemical, 165, 167
chemistry, 182
chemotaxis, 157, 161
chromaffin cells, 182
chromatin, 174, 179
chromium, 164
classes, 164
classical, 177
classification, 164
closure, 152, 159
clusters, 155
cobalt, 164, 182

collaboration, 179
communication, x, 151, 155
compensation, 167, 176
components, 152, 153, 154, 158, 165
compounds, x, 151
computer, 167
concentration, 152, 156, 157, 167, 169, 170, 172, 173, 176, 178, 182
condensation, 170, 174
conductance, 167
configuration, 167
confocal laser scanning microscope, 166, 180
control, 155, 156, 158, 167, 172, 173, 175, 176
control condition, 172, 175
coordination, 165
copper, 164
correlation, 178, 182
cortex, 157
cortical, 177, 181, 182
cortical neurons, 177, 181, 182
coupling, x, 151, 154, 155, 160, 162
culture, x, 163, 166, 167, 168, 172, 173, 181
culture media, 167
cytoplasm, 154, 158
cytosine, 166
cytoskeleton, 158, 161, 162
cytosolic, 180
cytotoxicity, 173

D

death, x, 163, 170, 173
defense, 165, 178
deficiency, 155, 161
definition, 177
delivery, 157
delta, 183
density, 164, 166
depolarization, 158, 165, 169, 176
depolymerization, 157, 158
destruction, 152
detachment, 172, 173
detection, 168
detoxification, 178
diatoms, 164, 183
dissociation, 152, 153
distilled water, 168
distribution, 159
drainage, 177
duration, 171
dyes, 180

E

electrical, 176, 179
emission, 169, 170
encoding, 166, 179
endogenous, x, 163
endothelial cell, 177, 181
endothelium, 177
engagement, 152
envelope, 168
environment, x, 163
enzymes, x, 152, 163
equilibrium, 164
ester, 156, 166
ethanol, 167
etiology, 164
evidence, 164, 165, 166, 180, 182
excitation, 168
exclusion, 167
exocytosis, 158, 162
exogenous, 182
experimental condition, 156, 157
exposure, x, 163, 169, 173, 174, 177, 178, 180

F

family, 154, 155, 160
fetal, 166
filters, 168
float, 172
flow, 167
fluorescence, 156, 162, 166, 169, 170, 181, 182
food, 164
fragmentation, 170

G

gastrointestinal, 165, 177
generation, 152, 153, 162
genes, 154
gentamicin, 166
gift, 166
glass, 166, 168
glial, 177
glutamate, 170
granules, 157, 162
graph, 171, 172, 173
graphite, 168
gravity, 167
growth, 179
guanine, 152

H

heavy metal, 164, 179, 181, 182
host, 152
human, 159, 160, 162, 164, 180, 181, 182
hypothesis, 154, 157

I

ice, 168
identification, 180
identity, 152, 177
images, 169
imaging, 180, 182
immersion, 166
in vitro, 166
in vivo, 182
inactivation, 170
incubation, 167, 172, 174, 175
indication, 178
indicators, 162, 166, 168, 174
induction, 165, 180
industrial, x, 163, 164
infection, 152
inflammatory disease, 158
inflammatory mediators, 152, 153
inhibition, 155, 157, 158, 160, 181
inhibitor, 152, 153, 157, 161
initiation, 174
injury, 165, 170, 178
inositol, x, 151, 152, 153, 155
insulinoma, 169, 170, 173, 180
integrin, 152, 159
integrity, 174
intensity, 166, 170, 178
interaction, 154, 162, 164
interactions, 182
interface, 167
intestine, 178, 179
intrinsic, 164
ionic, x, 163, 165, 179, 180
ions, x, 154, 156, 163, 164, 169, 182
iron, 164, 178, 179, 182
iron deficiency, 178

K

kidney, 177
killing, 152
kinetics, 177, 181
knockout, 181

L

lanthanum, 170, 182
latency, 178
lead, x, 163, 164, 165, 177, 179, 180, 181, 182
lipid, 170, 180
liposomes, 180
localization, 157, 158, 160
location, 157
luminal, 177
lymphocytes, 154, 160
lysine, 166

M

magnesium, 164
manganese, 164, 182
media, x, 151
membranes, x, 152, 153, 163, 166, 170, 177
mercury, 164
metabolism, 165, 180
metabolites, 152
metal content, 175
metal ions, x, 163, 168
metals, x, 163, 164, 165, 178, 180
mice, 182
micronutrients, 164
microphotographs, 171
microscopy, 158, 166, 169
migration, 157, 161
mimicking, 165
mimicry, x, 163, 178, 179, 180
misleading, 164
modulation, 172, 180
molecular structure, 154
molecules, 157, 159
molybdenum, 164
monolayer, 166, 171, 172
morphological, 172
morphology, 174
mortality, 170, 172, 173
mutation, 155, 161
myelin, 180

N

necrosis, 170, 174, 177, 180, 181
network, x, 151, 158
neurodegenerative, 164
neurodegenerative disease, 164
neuronal cells, 166

neurons, 165, 169, 170, 177, 180, 183
neutrophils, x, 151, 153, 154, 156, 157, 158, 159, 160, 162
nickel, 182
nimodipine, 167, 169, 170, 171, 173, 175, 176
norepinephrine, 182
normalization, 175
nuclear, 168, 174, 175, 179
nucleus, 174
nutrients, 164

O

obligate, 158
observations, 165
oil, 166
oocytes, 181
ovary, x, 163, 164, 166, 179
oxidative, 164, 179
oxygen, 152

P

particles, 152
passive, x, 163, 177, 178
pathways, x, xi, 152, 158, 161, 163, 164, 166, 169, 170, 178, 181
peptide, 182
perfusion, 180
permeability, 182
permeant, 166, 170
permeation, x, xi, 163, 164, 165, 169, 170, 177, 180, 181
phagocyte, 159
phagocytosis, 152, 159, 162
pheochromocytoma, 180
phone, 163
phosphatidylserine, 168, 174, 181
phospholipids, 174
phosphorylation, 152, 159, 162
photon, 180, 182
physiological, x, 163, 169, 177
pinhole, 166, 180
plasma, x, 151, 152, 153, 154, 155, 156, 157, 161, 162, 163, 165, 174, 180, 181, 182
platelets, 152, 154, 159, 160
play, 177
poisoning, 177
polarity, 157, 161
polymerization, 157
potassium, 164, 167
priming, 159

probe, 169
production, 152, 153, 158, 160
proliferation, 166
protection, 173
protective mechanisms, 165
protocol, 167, 171
proximal, 177
pulse, 172
pumps, x, 152, 163, 178

R

radius, 165
range, 170
rat, 170, 177, 179, 180, 181, 182
receptors, x, 151, 152, 153, 154, 155
recognition, 159
reconcile, 157
recovery, 173
redistribution, 155, 174, 181
reduction, 176, 180
regulation, x, 151, 154, 158, 178
regulators, 157
relationship, 153
remodeling, 179
research, 164
residues, 154, 155
resistance, 167, 181
respiratory, 160
reticulum, x, 151, 152, 154, 155, 156, 157, 158, 160
returns, 152
risk, 178
room temperature, 166

S

saline, 166, 169, 180
sample, 167, 168
saturation, 170
secretion, x, 151, 152, 162
segmentation, 168
sensitivity, xi, 164, 165, 172, 173, 177
serum, x, 163, 166, 167, 168, 172
shape, 172, 173
sharing, 155
signalling, 154, 158, 159, 161
signals, x, 151
similarity, 165
sites, x, 163, 165
smooth muscle, 166
sodium, 164, 180
software, 167

soil, 164
solutions, 167
somatic cell, 183
specialized cells, 164
species, x, 163, 165
specificity, x, 163
spectrum, x, 163
standard error, 168
stimulus, 169, 170, 178, 181
strategies, 178
stress, 180
substances, 164
substitution, 167
suffering, 172
susceptibility, 178
systems, x, 163, 165, 177, 178

T

targets, 158
threat, 164
time, x, 151, 152, 158, 159, 171, 172, 174, 178
tissue, 165
toxic, x, 163, 164, 165, 166, 168, 169, 170, 172, 177, 178, 179, 180
toxic effect, 165
toxic metals, x, 163, 164, 165, 166, 178, 179, 180
toxicity, x, 163, 164, 170, 175, 178, 180, 181
toxicological, 177
traffic, 157, 158
transduction, x, 151

translocation, x, 152, 159, 163, 174
transport, x, xi, 163, 164, 165, 170, 177, 178, 179, 180, 181
tyrosine, 162

U

uniform, 155, 157

V

variation, 174
visible, 168

W

wild type, x, 163, 166, 167, 171, 172, 173, 174, 175, 176

X

xenobiotic, x, 163

Z

zinc, 164, 180, 181, 182